LIST OF

Already

A Biochemi
   Nutrition   iggs
Biochemical Genetics (seco
Biological Energy Conserva
   (second edition)
Biomechani
Brain Bioc
Cellular De
Cellular De
Cellular Re
Control of
   (second edition)   P. Cohen
Cytogenetics of
   Animals
Enzyme Kine
Functions of Bio
Genetic Engineering: Cloning DNA   Glover
Hormone Action   A. Malkinson
Human Evolution   B.A. Wood
Human Genetics   J.H. Edwards
Immunochemistry   M.W. Steward
Insect Biochemistry   H.H. Rees
Isoenzymes   C.C. Rider, C.B. Taylor
Metabolic Regulation   R. Denton, C.I. Pogson
Metals in Biochemistry   P.M. Harrison, R. Hoare
Molecular Virology   T.H. Pennington, D.A. Ritchie
Motility of Living Cells   P. Cappuccinelli
Plant Cytogenetics   D.M. Moore
Polysaccharide Shapes   D.A. Rees
Population Genetics   L.M. Cook
Protein Biosynthesis   A.E. Smith
RNA Biosynthesis   R.H. Burdon
The Selectivity of Drugs   A. Albert
Transport Phenomena in Plants   D.A. Baker

Membrane Biochemistry   E. Sim
Muscle Contraction   C. Bagshaw
Glycoproteins   R.C. Hughes
The Biochemistry of Membrane
   Transport   I.C. West

Editors' Foreword

The student of biological science in his final years as an undergraduate and his first years as a graduate is expected to gain some familiarity with current research at the frontiers of his discipline. New research work is published in a perplexing diversity of publications and is inevitably concerned with the minutiae of the subject. The sheer number of research journals and papers also causes confusion and difficulties of assimilation. Review articles usually presuppose a background knowledge of the field and are inevitably rather restricted in scope. There is thus a need for short but authoritative introductions to those areas of modern biological research which are either not dealt with in standard introductory textbooks or are not dealt with in sufficient detail to enable the student to go on from them to read scholarly reviews with profit. This series of books is designed to satisfy this need. The authors have been asked to produce a brief outline of their subject assuming that their readers will have read and remembered much of a standard introductory textbook of biology. This outline then sets out to provide by building on this basis, the conceptual framework within which modern research work is progressing and aims to give the reader an indication of the problems, both conceptual and practical, which must be overcome if progress is to be maintained. We hope that students will go on to read the more detailed reviews and articles to which reference is made with a greater insight and understanding of how they fit into the overall scheme of modern research effort and may thus be helped to choose where to make their own contribution to this effort. These books are guidebooks, not textbooks. Modern research pays scant regard for the academic divisions into which biological teaching and introductory textbooks must, to a certain extent, be divided. We have thus concentrated in this series on providing guides to those areas which fall between, or which involve, several different academic disciplines. It is here that the gap between the textbook and the research paper is widest and where the need for guidance is greatest. In so doing we hope to have extended or supplemented but not supplanted main texts, and to have given students assistance in seeing how modern biological research is progressing, while at the same time providing a foundation for self help in the achievement of successful examination results.

General Editors:

**W.J. Brammar, Professor of Biochemistry,
University of Leicester, UK**

**M. Edidin, Professor of Biology,
Johns Hopkins University, Baltimore, USA**

Antibodies:
Their structure and function

M.W. Steward
Professor of Immunology
London School of Hygiene and Tropical Medicine

Chapman and Hall
London and New York

First published 1984
by Chapman and Hall Ltd
11 New Fetter Lane, London EC4P 4EE
Published in the USA
by Chapman and Hall
733 Third Avenue, New York NY 10017

© *1984 M.W. Steward*

Printed in Great Britain by
J. W. Arrowsmith Ltd, Bristol

ISBN 0 412 25640 1

British Library Cataloguing in Publication Data

Steward, M.W.
    Antibodies.–(Outline studies in biology)
    1. Immunoglobulins
    I. Title   II. Series
    574.2'93   QR186

    ISBN 0-412-25640-1

Library of Congress Cataloging in Publication Data

Steward, M.W. (Michael W.),   1940-
    Antibodies, their structure and function.

    Includes bibliographies and index
    1. Immunoglobulins. I. Title. (DNLM: 1. Antibodies
-- Physiology. QW 575 S849A)
QR186. 7S74 1984 616.07'93 83-24035
ISBN 0-412-25640-1 (Pbk)

Contents

1	**Introduction**	**7**
2	**Isolation and purification of immunoglobulins and specific antibodies**	**12**
2 1	Induction of serum antibodies	12
2.2	Isolation of immunoglobulins	13
2.3	Isolation of specific antibodies	15
2.4	Monoclonal antibodies produced by hybrid myelomas (hybridomas)	15
	References	20
3	**General structure of immunoglobulins**	**21**
3.1	The basic four chain model for IgG	21
3.2	Immunoglobulins are glycoproteins	23
3.3	Amino acid sequence studies	23
3.4	The antibody binding site	26
3.5	Immunoglobulin domains	28
3.6	Allotypes and idiotypes of immunoglobulins	29
	References	35
4	**Antibody–antigen interaction**	**37**
4.1	The intermolecular forces involved in antibody–antigen interactions	37
4.2	The measurement of antibody–antigen reactions	39
4.3	The study of the chemistry of antibody–antigen reactions	41
4.4	The thermodynamics of antibody–antigen reactions	43
4.5	The kinetics of the antibody–antigen reactions	54
4.6	The biological aspects of antibody affinity	55
4.7	The specificity and cross-reactivity of antibody–antigen interactions	61
	References	65
5	**Structure and biological activities of the immunoglobulin classes**	**67**
5.1	Immunoglobulin G	69
5.2	Immunoglobulin A	71
5.3	Immunoglobulin M	75
5.4	Immunoglobulin E	78
5.5	Immunoglobulin D	79
5.6	Effector functions of antibodies	80
	References	84

6	**The control of antibody production**	**85**
6.1	The synthesis and secretion of antibody molecules	85
6.2	The genetic control of antibody biosynthesis	86
6.3	The generation of antibody diversity	93
	References	94
Index		**95**

1 Introduction

It has been appreciated for a very long time that individuals who recover from an infectious disease rarely contract the disease again. In describing a plague in Athens some 2500 years ago, Thucydides made the point that it was known that those who had recovered from the plague could tend the sick without fear of catching the disease again – in other words, they were *immune*. Over the following centuries, many attempts have been made to induce this state of immunity or resistance to disease, among these the Chinese custom of inhalation of the crusts of smallpox lesions for protection of the disease. In addition, inoculation of healthy individuals with material from the lesions – a process called *variolation* – was also used. However, it was the work of Jenner which revolutionized the approach to inducing immunity. In his famous report in 1798 he recorded how inoculation with cowpox (vaccinia) leads to the development of immunity to smallpox and from which is derived the term *vaccination*. Subsequent work showed that immunity to infectious diseases could be induced by injection of products of the organism. Thus von Behring in 1888 demonstrated that non-lethal doses of the toxin from diphtheria induced immunity to the organism. It was work on toxin-induced immunity which lead to the demonstration by von Behring and Kitasato in 1890 that immunity to tetanus following injection of tetanus toxin was the result of the appearance of a 'factor' in the serum which neutralized the toxin. This anti-toxin activity could be transferred to normal animals by injecting serum from immune animals. It thus appeared that the body is able to respond to infectious organisms or injurious substances by producing serum components which would combine with these agents and neutralize their effects. These components were thus called '*antibodies*' and the agents provoking their production were termed '*antigens*'. In the decade or so following these early observations it was demonstrated that antibodies could specifically lyse bacteria (bacteriolysis), precipitate bacteria and agglutinate bacteria (cause them to clump together). Furthermore, it was shown that antibodies could be produced against toxins other than those from bacteria and against non-toxic substances such as proteins.

The reactions of antibodies and antigens attracted the attention of eminent scientists such as Paul Ehrlich who was the first person to quantitatively study the precipitation reaction of anti-toxin and toxin. The great physical chemist Arrhenius also became interested in immunological precipitation reactions and coined the term 'immunochemistry'. He stated:

'I have given to these lectures the title 'Immunochemistry' and wish with this word to indicate that the chemical reactions of substances that are produced by the injection of foreign substances into the blood of animals, i.e. by immunization, are under discussion in these pages.

From this it follows also, that the substances with which these products react as proteins and ferments, are to be here considered with respect to their chemical properties.'

Our current concepts have, of course, broadened from those of the early immunochemists but this definition is still applicable to modern immunochemistry and could be rephrased in modern terms as the study of the chemistry of antibodies and antigens and of their mechanism of interaction.

These studies thus pointed clearly to the importance of antibodies in the immune response to infections. However, at the time these studies on serum antibodies were proceeding, Metchnikoff (1882) produced evidence that he strongly used to support the view that *phagocytosis* ('cellular eating') of microorganisms by leucocytes was the major host defence against infection. His views were bitterly opposed by the proponents of the antibody (humoral) hypothesis of host protection. Indeed, it was not until the work of Sir Almroth Wright in 1903, who showed that antibodies actually aided the phagocytosis of bacteria (a process which he called *opsonization*) that these two hypotheses of immunity were reconciled. Thus both cells and antibody were shown to be necessary for an effective host response to infection. The purpose of this book, however, is to focus on our understanding of the nature and function of antibodies, but in doing so it is not to suggest that cell-mediated responses are not important!

Although the presence of antibodies in the serum of immunized animals was recognized in the late 19th century, as discussed above, it was not until 1938 that significant advances in the knowledge of their chemical nature were made. At this time, Tiselius and Kabat demonstrated that the antibody activity of an <u>antiserum</u> is associated with the γ-globulin fraction of the serum. The technique employed was that of electrophoresis which Tiselius had developed in 1937. The result of electrophoresis of hyperimmune anti-ovalbumin serum is shown in Fig. 1.1 in which an increased level of γ-globulin can be seen. The electrophoresis pattern of the same serum after

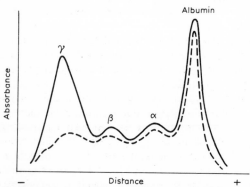

Fig. 1.1 Electrophoresis of anti-ovalbumin antiserum before and after specific precipitation of the anti-ovalbumin antibodies with the antigen. ———, anti-ovalbumin before precipitation with antigen; – – –, anti-ovalbumin after precipitation with antigen [1].

the removal of the specific antibodies by precipitation with the antigen-ovalbumin is also shown. From this, it is clear that the specific precipitation with antigen removed a significant amount of the γ-globulin, from which it was concluded that the antibodies to ovalbumin were γ-globulins. Globulins which function as antibodies are now termed *immunoglobulins*. One characteristic feature of antibodies observed from these and subsequent studies was their heterogeneity, particularly in terms of their electrical charge (and amino acid sequence) (Fig. 1.2).

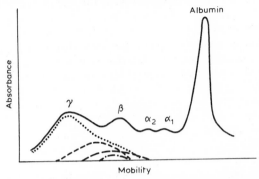

Fig. 1.2 The electrophoretic mobility of immunoglobulin classes. ———, whole serum; ...., IgG; — — — —, IgA; — —, IgM; — · — · —, IgD.

As a result of these observations, it has become clear that immuno-globulins can be divided into various classes on the basis of structural differences, not related to their specificity for antigen. Therefore, in man, five classes of immunoglobulins have been characterized and have been termed immunoglobulin G(IgG), IgM, IgA, IgD and IgE. Similar classes of antibody exist in other species. The predominant immunoglobulin in serum is IgG which has an approximate molecular weight of 150 000 and a sedimentation coefficient of 7S in the ultracentrifuge. The detailed structural characteristics of this and the other classes will be described in detail later in this book.

This heterogeneity in structural and functional terms, has made the structural study of antibodies difficult. However, the occurrence of homo-geneous immunoglobulins in the blood and urine of patients with myelo-matosis and macroglobulinaemia has provided a useful source of large amounts of antibody for the study of immunoglobulin structure. These antibodies are the products of antibody-forming cell tumours (myelomas) and are called myeloma proteins. They derive from a single cell (clone) and thus all the proteins produced by this tumour are homogeneous or mono-clonal. Fig. 1.3 represents examples of cellulose acetate electrophoresis profiles of normal human serum (A), hyperimmune serum (B) and three different sera from patients with myeloma (C–E). Myeloma tumours have also been obtained in mice and by serial passage have provided a continuous source of homogeneous antibodies. In addition, the intensive immunization of rabbits with streptococci or pneumococci over long periods results in

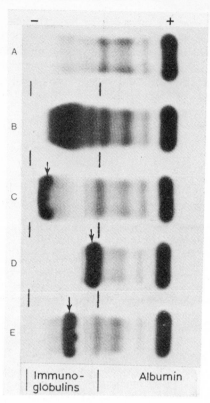

Fig. 1.3 Cellulose acetate electrophoresis of five human sera. Serum A is a normal serum; serum B has elevated levels of polyclonal, heterogeneous antibody; sera C—E are from patients with myelomatosis and exhibit monoclonal proteins (arrowed) with different electrophoretic mobilities (after [2]).

the production, in a proportion of animals, of antibodies of restricted heterogeneity, which in some are virtually monoclonal (Fig. 1.4).

More recently a technique has been described [4] by which it is possible to produce unlimited amounts of homogeneous, monoclonal antibodies with antibody specificity against any desired antigen. The monoclonal antibodies are produced by somatic cell hybrids, called *hybridomas*, and the antibodies are specific for an individual antigenic determinant on the antigen molecule. By utilizing this approach, it is now possible to produce sufficient amounts of monoclonal antibodies for structural studies. Details of the technique and its potential are discussed in Chapter 4.

The information about the structure and function of antibodies which will be discussed in this book was derived from studies on antibodies obtained from the sources described above.

References

[1] Tiselius, A. and Kabat, E. (1939), *J. Exp. Med.*, **69**, 119.

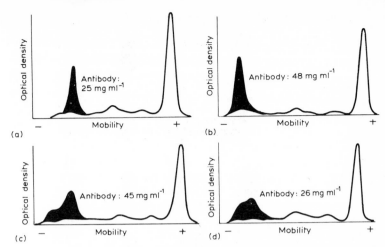

Fig. 1.4 Cellulose acetate electrophoresis of sera from rabbits immunized with Group C streptococci. The shaded areas represent antibody precipitable by the group C carbohydrate. Patterns (a) and (b), antibodies of restricted heterogeneity and patterns (c) and (d), heterogeneous antibodies (after [3] with permission).

[2] Eisen, H. B. (1980), *Immunology. An introduction to molecular and cellular principles of the immune response*, 2nd edn, Harper and Row, Hagerstown, Maryland.
[3] Krause, R. (1971), in *Immunoglobulins* (eds S. Kochura and H. G. Kunkel), *Ann. NY Acad. Sci.*, **190**, 276.
[4] Kohler, G. and Milstein, C. (1975), *Nature*, **256**, 495.

2 Isolation and purification of immunoglobulins and specific antibodies

2.1 Induction of serum antibodies

When an animal encounters an antigen either by infection or by deliberate injection, antibodies are produced which appear in the serum. In order to study the chemical and physico-chemical properties of these highly heterogeneous protein molecules, adequate amounts need to be isolated and purified. However, their heterogeneity of size, structure, charge and biological activity has made isolation and purification a challenging problem for immunochemists.

Immunoglobulins of a vast range of specificities for antigen exist of course in the serum of unimmunized animals and can be isolated by the techniques described below. In order to isolate and study specific anti-bodies, significant levels of these antigen-specific immunoglobulin molecules need to be induced by appropriate immunization procedures. When an animal is first injected with an antigen, for example tetanus toxoid, there is a period of several days when no antibody can be detected in the blood. After a period of 7–10 days, antibodies then appear and reach a peak at about 14–21 days after which the levels fall (Fig. 2.1). This first response to the antigen is termed the *primary* antibody response. If, after a period, the animal is given a second injection of the antigen, the levels of antibody in the blood begin to rise within two or three days and subsequently reach levels in excess of those achieved in the primary response. This is termed the *secondary* response. This accelerated response is the result of the stimulation by the antigen of antibody-forming cells of the immune system which had been sensitized by the earlier exposure to antigen. Generally the levels of antibody in the secondary response remain elevated for several weeks. Antibodies in the primary response are generally the IgM class (19S in the ultracentrifuge) whilst those in the secondary are of the IgG

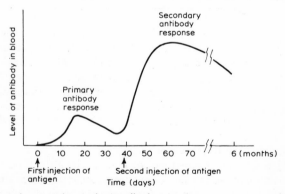

Fig. 2.1 The primary and secondary antibody response.

12

class (7S in the ultracentrifuge). Thus in order to obtain specific antibody, at least two injections of antigen are required. However, many antigens are poor at eliciting an antibody response when injected in saline, that is, they are poor immunogens. This poor immunogenicity can be overcome by aggregating the antigen or more often, by the use of immunological adjuvants. These are substances which increase the level of antibody response when injected with the antigen. A range of adjuvants are available (Table 2.1). The most widely used being the mineral oil adjuvants which contain an emulsifier as well as the oil. Injection of the antigen as a water-in-oil emulsion stimulates a long-lived antibody response. The precise mechanism by which this is achieved is unclear but may be related to the slow release of the antigen from the emulsion 'depot' into the animal (rather like a long series of injections of small amounts of antigen). The most well-known adjuvants of this type are Freund's incomplete adjuvant and Freund's complete adjuvant, which contains heat-killed *Mycobacterium tuberculosis* organisms in addition to the oil.

Table 2.1 Commonly-used adjuvants

Water-in-oil

    Freund's incomplete adjuvant
    Freund's complete adjuvant

Mineral

    Antigen absorbed onto aluminium hydroxide
    Quarternary ammonium salts

Bacterial

    Bacille Calmette Guerin (BCG) organisms
    Pertussis organisms
    Corynebacterium parvum organisms

Bacterial products

    Endotoxin

Polynucleotide

    Poly I-poly C

The immunization protocol used for the generation of high levels of antibody will depend upon a number of factors including dose of antigen, nature of adjuvant, route of infection (intraperitoneal, intradermal, subcutaneous, etc.) and the responsiveness of the animal. This process tends to be more of an 'art' than science. However, having successfully induced appropriate levels of antibody, the serum can be obtained and the specific antibody isolated from the other proteins in the serum.

2.2 Isolation of immunoglobulins

Immunoglobulins are, fortunately for immunochemists, robust molecules and they can survive in a variety of environments including heating to $56°C$ and storage at room temperature for extended periods. They can maintain their antibody activity after treatment with high or low pH for short periods and even after being in contact with detergents or urea. Methods which have been developed for the isolation of immunoglobulins from other plasma proteins (and from each other) make use of their basic nature, differing solubilities in various solvents and their relatively high isoelectric points (pI for IgG = 5.8–7.3). Examples of the techniques employed are shown in Table 2.2 and detailed information can be obtained from references [1–5]. The use of any one of these methods by itself will not yield a pure immunoglobulin preparation and in general, a combination of techniques needs to be employed.

Fractional precipitation techniques are commonly used to effect a concentration of crude immunoglobulin from serum, and simple dialysis of serum against water to produce the insoluble epglobulin and soluble pseudoglobulin fractions may be used as a first step in the isolation of immunoglobulin (IgM). However, it has been appreciated for over one hundred years that mineral salts can be used to precipitate or fractionate proteins and ammonium, sodium and magnesium sulphates have been utilized. As the concentration of salt increases, the protein molecules become less hydrophilic, and the subsequent hydrophobic interaction

Table 2.2 Methods for immunoglobulin isolation

Method	Examples
Fractional precipitation	
(i) Neutral salts	Ammonium sulphate
	Sodium sulphate
	Magnesium sulphate
(ii) Organic solvents	Ethanol
(iii) Metal ions	Zinc
(iv) Organic cations	Rivanol (2 ethoxy-6,9 diaminoacridine lactate)
Electrophoretic separation	
(i) Carrier free media	Free boundary electrophoresis
(ii) Solid support	Zone electrophoresis on cellulose, acrylamide, starch gel Pevikon
(iii) Isoelectric focusing	Liquid media − sucrose gradient
	Gel media − polyacrylamide
Ion-exchange chromatography	
(i) Anion exchangers	Aminoethyl (AE−)
	Diethylaminoethyl (DEAE−) and quarternary amino ethyl (QAE−) cellulose, Sephadex, Sepharose
(ii) Cation exchangers	Carboxymethyl (CM−) cellulose, Sephadex, Sepharose
Gel filtration	Sephadex G-200, Sepharose 2B, 4B, 6B
Preparative ultracentrifugation	Sucrose density gradients
	Salt gradients

between the molecules eventually leads to their precipitation. The basis of fractional precipitation is the fact that precipitation of different proteins occurs at different salt concentrations. Thus at certain concentrations of ammonium sulphate, (i.e. $1.2-1.8$ M) immunoglobulins are precipitated but other proteins, such as albumin, remain in solution. This procedure therefore provides a useful first stage in the isolation of immunoglobulins. Further purification can be achieved by employing other techniques shown in Table 2.2. Methods are available for the isolation of each of the five immunoglobulin classes (IgG, IgA, IgM, IgE and IgD) which exploit the chemical and physico-chemical differences between molecules of the various classes. Examples of these methods are given in Table 2.3 which is given as an illustration only and does not claim to be fully comprehensive. A protocol for the purification of IgG, IgM and IgA from a human plasma sample is outlined in Fig. 2.2.

2.3 Isolation of specific antibodies

One of the most widely used methods of isolating pure antibodies is that of affinity chromatography utilizing an immunoadsorbent, which is essentially an insolubilized form of the antigen formed by coupling the antigen to an insoluble particle. Commonly, antigens are covalently bound to substances such as cyanogen-bromide-activated sepharose. The principle of affinity chromatography is shown in Fig. 2.3. The impure antibody is reacted with the insolubilized antigen to form an antigen-antibody complex and all material not bound is washed away. Since antibody–antigen interactions are based on non-covalent bonds (see Chapter 4), the complex of antibody and its corresponding antigen can be relatively easily broken by agents such as acid buffers (pH 2.0), dilute ammonia or chaotropic reagents (thiocyanate, bromide or iodide), yielding a pure preparation of antibody. Whilst this technique has been widely used, certain problems with it do exist, including the denaturation of the antibody by the dissociating agents and the possibility that antibody of the highest affinity (see Chapter 4) for the antigen may not in fact be readily eluted from the immunoadsorbent. Nevertheless, this technique does provide a convenient method for the purification of specific antibodies.

2.4 Monoclonal antibodies produced by hybrid myelomas (hybridomas)

As discussed earlier, myelomas have provided an invaluable source of monoclonal immunoglobulin molecules for structural studies. One major limitation of these molecules is that in most cases we have no real idea as to the nature of the antigen to which the antibody is directed. Methods for determining the specificity of monoclonal antibodies have been devised but they are time-consuming. One of the most exciting developments in immunology for many years was the description by Kohler and Milstein in 1975 [7] of a general method for the production of large quantities of monoclonal antibodies. The general principle of the production of an antibody-producing hybridoma is shown in Fig. 2.4 and consists of the fusion of a non-secreting myeloma tumour cell (with approximately 65 chromosomes) and an antibody-producing B-cell (with the normal comple-

15

Table 2.3 Outline of techniques for preparation of immunoglobulin classes

Immunoglobulin class	Source	Initial precipitation	Chromatographic Steps	Other procedures
IgG	Normal serum	40% $(NH_4)_2SO_4$ or Na_2SO_4	(i) DEAE-cellulose (ii) Sephadex G200	Protein A (binds IgG)
IgM	Normal serum; Waldenstrom's macroglobulinaemia serum	Dialysis v. water (use euglobulin) or 45% $(NH_4)SO_4$ or 6% polyethylene glycol	Sepharose 4B or 6B	Protein A (to remove IgG) Protamine sulphate (binds IgM)
IgA (serum)	Normal serum; Myeloma serum	(i) Dialysis vs. water (use pseudoglobulin) (ii) Zinc sulphate (to precipitate IgG)	(i) DEAE-cellulose (ii) Sephadex G200	Affinity chromatography using insolubilized antibodies to other Ig classes
IgA (secretory)	Colostrum, milk	(i) Delipidation (ii) Decaseination	(i) Sephadex G200 (ii) DEAE-cellulose	
IgE	Myeloma serum; ?Atopic serum	18% Na_2SO_4	(i) DEAE-cellulose (ii) Sephadex G200	Insolubilized antibodies to other classes
IgD	Normal serum; Myeloma serum	45% $(NH_4)_2SO_4$	(i) DEAE-cellulose (ii) Sephadex G200	Insolubilized antibodies to other classes

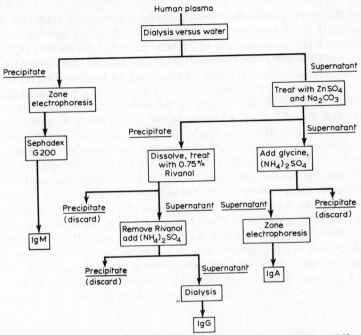

Fig. 2.2 The purification of IgM, IgG, and IgA from human plasma (from [6] with permission).

ment of 40 chromosomes). The fused cells retain the 'immortality' of the myeloma cell line and also continue to secrete the antibody produced by the B cell. They possess fewer chromosomes than the sum of the two cells being fused, but have greater than twice the number of chromosomes in normal mouse cells. The fused cells (hybridoma) may then be maintained in tissue culture or they may be passaged in animals as tumours where they produce large amounts of antibody in ascites fluid. It was soon realized

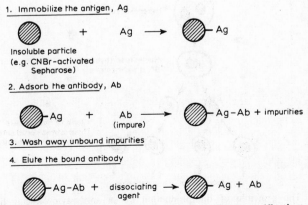

Fig. 2.3 The principle of affinity chromatography for antibody purification.

17

that this method could be utilized to produce monoclonal antibodies with specificity for any antigen. The important details of the technique are as follows: B-cell myeloma cells which lack the purine salvage enzyme hypoxanthine guanosine phosphoribosyl transferase (HGPRT) are selected for the fusion. This selection is achieved by exploiting the fact that HGPRT⁻ cells are able to grow in a medium containing 8-azaquanine whereas HGRPT⁺ cells die in this medium. Following the hybridization of these myeloma cells and the spleen cells (from an animal appropriately immunized with the desired antigen) with polyethylene glycol, the cells are cultured in a selective medium containing hypoxanthine, aminopterin and thymidine (HAT). In this medium, the tumour cells are killed, the normal, non-fused cells die after a short period in culture and the hybridomas survive. The reasons for the selective effect of the HAT medium will be briefly described. Cells have two ways of producing nucleic acid: (a) *de novo* synthesis and (b) the salvage pathway in which nucleotides from degraded nucleic acids are utilized by a process which requires the enzyme HGPRT. In the presence of aminopterin, *de novo* nucleic acid synthesis is blocked but normal cells can survive by using the salvage pathway and the nucleotides in the HAT medium. However, the HGPRT⁻ tumour cells die in the presence of aminopterin because they can use neither *de novo* synthesis nor the salvage

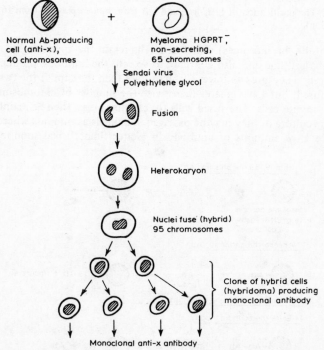

Fig. 2.4 The principle of the production of an antibody-producing hybridoma (after [8]).

pathway. The hybridoma cells survive because they have HGPRT derived from genes received from the normal spleen-cell parents.

The hybridomas which grow are screened for their production of antibody by appropriate sensitive assays e.g. ELISA or radioimmunoassay (see Chapter 4) and the antibody-producing hybridomas are cloned. This is a process in which an individual antibody producing cell is isolated and then grown into a homogeneous population (clone) of cells. Two widely-used methods used to achieve this are (a) cloning in soft agar and (b) cloning by limiting dilution. In the former, high dilutions of the cells are made in agar, and appropriate colonies are picked out from the soft agar and grown in culture. In the latter, high dilutions of the cells are made so that individual cells can be transferred to tissue culture medium in microtitre plates and grown. Wells containing one colony are then grown in culture. An outline of the experimental protocol for the production of an antibody-producing hybridoma is shown in Fig. 2.5. Hybridomas can be grown in tissue culture where up to $10\,\mu g\,ml^{-1}$ of specific antibody may be obtained. However, they will also grow in the peritoneal cavity of animals of the same strain (histocompatibility type) as the tumour cell donor and the

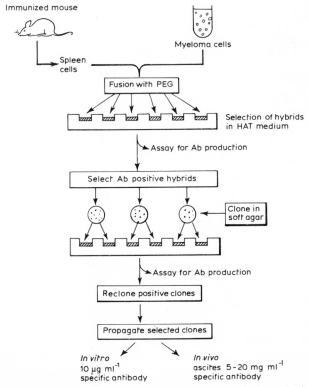

Fig. 2.5 Experimental protocol for the production of an antibody-producing hybridoma (from [9] with permission).

spleen cell donor. The hybridomas then secrete large amounts of antibody $(5-20 \text{ mg ml}^{-1})$ into the ascitic fluid. The latter is clearly a very attractive method but the ascites fluid will of course contain other proteins and immunoglobulins from the host. The monoclonal antibody may, however, be purified by utilizing the techniques described earlier, in particular, affinity chromatography with the appropriate insolubilized antigen.

References

[1] Williams, C. A. and Chase, M. A. (eds) (1967), *Methods in Immunology and Immunochemistry* (Vols I and II), Academic Press, London.
[2] Schultze, H. E. and Heremans, J. F. (1960), *Molecular Biology of Human Proteins* (Vol. I), Elsevier, Amsterdam.
[3] Weir, D. M. (ed.) (1978), *Handbook of Experimental Immunology*, 3rd edn, Blackwell Scientific Publications, Oxford.
[4] Johnstone, A. and Thorpe, R. (1982), *Immunochemistry in Practice*, Blackwell Scientific Publications, Oxford.
[5] Hudson, L. and Hay, F. C. (1980), *Practical Immunology*, 2nd edn, Blackwell Scientific Publications, Oxford.
[6] Schwick, H. G., Fischer, J. and Geiger, H. (1968), *Progress in immunobiological standard*, **4**, 96.
[7] Kohler, G. and Milstein, C. (1975), Nature, **256**, 495.
[8] Eisen, H. N. (1980), *Immunology. An Introduction to molecular and cellular principles of immune responses*, Harper and Row, Hagerstown, Maryland, p. 340.
[9] Secher, D. S. (1980), *Immunology Today*, **1**, 23.

3 General structure of immunoglobulins

Three groups of antibodies can be detected in serum by ultracentrifugation: (a) proteins of molecular weight approximately 150 000 and a sedimentation coefficient of 7S, (b) proteins of molecular weight approximately 300 000 (9–11S) and (c) proteins of molecular weight approximately 900 000 (19S). The 7S is the major group and consists predominantly of IgG. This immunoglobulin accounts for about 70% of the total serum immunoglobulins in human serum.

3.1 The basic four chain model for IgG

The first attempts to investigate the structure of antibodies were made by R. R. Porter [1] who demonstrated that when 7S rabbit antibodies were incubated with the proteolytic enzyme papain in the presence of cysteine, three major fragments I, II and III were obtained. Using carboxymethyl-cellulose ion exchange chromatography to separate the fragments, Porter showed that one of these (fragment III) could be crystallized (subsequently called Fc 'fragment crystallizable') and that fragments I and II were identical to each other and unlike the Fc, were able to combine with the antigen. These observations thus accounted for the known valency of two for IgG antibodies and these fragments were subsequently termed Fab ('fragment antigen binding'). Nisonoff and his co-workers [2] showed that on treatment of 7S antibody with pepsin, a bivalent 5S antibody fragment is produced which, when reduced yields two monovalent 3.5S fragments. Further studies by G. M. Edelman showed that the constituent chains of immunoglobulin can be isolated by reduction of the interchain disulphide bonds with sulphydryl reagents in urea solution or by reduction followed by alkylation of the free –SH groups and subsequent dissociation of non-covalent bonds by gel filtration in acid [3]. In recognition of this work, Porter and Edelman shared the Nobel Prize for Medicine in 1972. These observations formed the basis on which Porter [4, 5] postulated a four polypeptide chain structure for IgG (Fig. 3.1). This fragmentation of IgG by enzyme and chemical treatment is shown diagrammatically in Fig. 3.2.

Thus IgG consists of two heavy and two light chains linked by covalent disulphide bonds and other non-covalent forces (e.g. hydrogen bonding). The heavy chains of human IgG with 450 amino acid residues (M_r 50 000) are approximately twice the size of the light chains which have 220 amino acid residues (M_r 23 000). This general structure for IgG has been confirmed by electron microscopy of antibody to DNP complexed to bis-N-DNP-octamethylene diamine which is a bifunctional hapten molecule consisting of an eight-carbon chain with a dinitrophenol group at each end. In the presence of this reagent, anti-DNP antibodies appear by negative staining as three molecules linked by the (invisible) reagent via their Fab regions

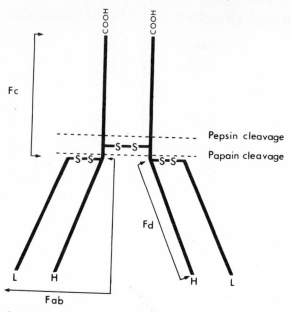

Fig. 3.1 The basic four chain structure for rabbit IgG as proposed by Porter.

with the Fc parts of the molecule forming projections at the corner of the triangle thus formed (Fig. 3.3).

Digestion with pepsin removes the Fc from the corner of the structure. It was concluded from these studies that IgG is a flexible Y-shaped molecule made up of three globular regions representing the Fab and Fc fragments. The 'arms' of these Y-shaped molecules (the Fab regions) can swing out to an angle of greater than $100°$ (Fig. 3.4(a)) because of the nature of the enzyme-sensitive region around the interchain disulphide bond – the so-called 'hinge' region. This flexibility accounts for the ability of the molecule to bind to two sites on a single particle (e.g. bacterium or virus) or to link two particles (Fig. 3.4(b), (c)) and arises as a result of the presence of a large number of proline residues. In addition, the loose folding of this region explains its susceptibility to proteolytic enzyme activity.

It has been shown that the four chain structure is basic for all immunoglobulin classes and that although the light chains are similar in each case, the heavy chains are specific for each class. Thus γ heavy chains are specific for IgG, α for IgA, μ for IgM, and δ for IgD and ϵ for IgE. Furthermore, immunoglobulins in all classes of vertebrates from fish to mammals have this basic four chain unit.

Light chains have been serologically separated into kappa (κ) and lambda (λ) types. It has been shown that in the human, 65% of immunoglobulins have κ light chains, the remainder having the λ type. In any one immunoglobulin molecule the two light chains are always of the same type and both κ and λ types can be associated with any heavy chain class. The

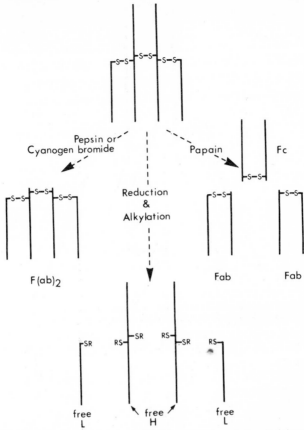

Fig. 3.2 The preparation of fragments of IgG by enzymic and chemical degradation.

$\kappa:\lambda$ ratio varies from immunoglobulin class to class and from subclass to subclass. For human immunoglobulins the overall ratio is 6:4. It should be mentioned that the $\kappa:\lambda$ ratio also varies from species to species.

3.2 Immunoglobulins are glycoproteins

At this point it should be mentioned that all immunoglobulins are glyco-proteins and the percentage of carbohydrate ranges from 2%–3% for IgG to as high as 14% for IgD. The carbohydrate is usually associated with the heavy chains (constant regions, see below) via covalent attachment to asparagine residues (and occasionally serine or threonine). Three functions have been ascribed to this carbohydrate: (a) to facilitate secretion from the antibody-synthesizing plasma cell; (b) to enhance the solubility of the immunoglobulin and (c) to protect the molecule from degradation.

3.3 Amino acid sequence studies

The availability of homogeneous immunoglobulins from sera of patients with myelomatosis has enabled structural studies at the amino acid sequence

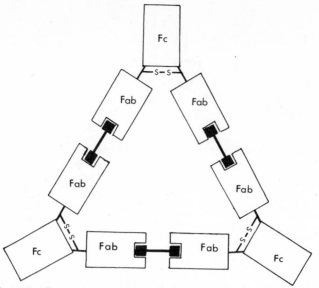

Fig. 3.3 A diagrammatic representation of a DNP–anti-DNP immune complex as seen in the electron microscope, showing the Y-shaped structure of the antibody. The antigen (■——■) is a bifunctional DNP hapten with the structure:

(after [6]).

level to be carried out. After IgG was shown to consist of light and heavy polypeptide chains which could be easily separated, it was demonstrated that the Bence-Jones proteins found in the urine of many patients with myelomatosis were very similar to the light chains of the patient's myeloma IgG. Several of these proteins have been isolated and their amino acid sequences determined. Comparison of the results obtained from these proteins have shown that the L-chains can be divided into two regions: the V or variable regions extending from amino acid positions 1–107 where there is a wide variability in amino acid sequence and the C or constant region extending from position 108–214 where the sequences are remarkably similar. A schematic representation of the constant and variable regions of human κ-Bence Jones proteins is shown in Fig. 3.5. Closed circles represent positions where different amino acids have been found in other κ-Bence Jones proteins and open circles represent amino acid which are constant. There is less sequence data on heavy chains but these also have constant and variable regions. The constant region accounts for approximately 75% of the chain and the variable region the remainder.

The variability observed in the amino acid sequences of the N-terminal regions of both heavy (V_H) and light chains (V_L) is not evenly distributed. Certain amino acid positions show exceptional degrees of variability

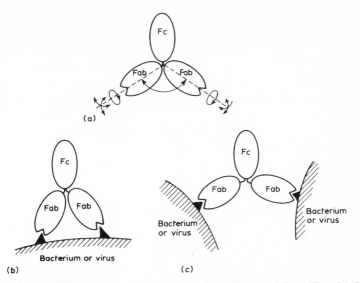

Fig. 3.4 The flexibility of the hinge region of IgG. (a) Segmental flexibility of IgG. The Fab can move in the directions shown by the arrows and the angle α between the two Fab arms is likely to be larger than 100° (after [7]); (b) monogamous bivalent binding to two determinants on a single particle; (c) bivalent binding of two particles via hinge region flexibility.

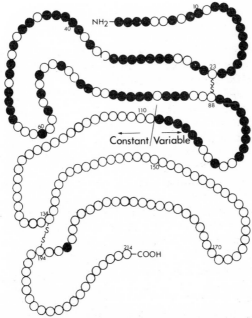

Fig. 3.5 The constant and variable regions of human κ myeloma light chains. ($\circ$), constant amino acid residues; ($\bullet$), variable amino acid residues (after [8]).

and have been termed the hypervariable regions or complementarity-determining regions (CDR) and are directly involved in the formation of the antigen-binding site. These regions are present in V_H, V_κ and V_λ and are shown in Fig. 3.6 from which it can be seen that there are three CDR in light chains and four in heavy chains [9]. It should be noted that adjacent to the CDR are regions of restricted variability which are termed the 'framework' residues, thought to provide the necessary rigid structures near which the CDR are sited. However, the hypervariable region represented by residues 82–89 of the heavy chain does not seem to play a part in the formation of the antigen-binding cleft.

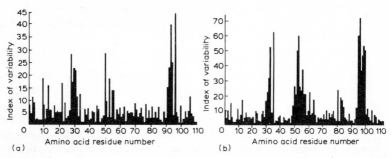

Fig. 3.6 Plots of amino acid sequences in the variable regions of both heavy and light chains illustrating the hypervariable or complementarity-determining regions; (a) human light chains; (b) human heavy chains [9].

When the amino acid sequences of a number of kappa or lambda human light chains were analysed it became apparent that the variable region sequences of individual light chains were closely related to those of other light chains (i.e. showed restricted variability). Thus on the basis of such homology the variable regions of κ and λ chains can be divided into subgroups. In man there are four κ subgroups (designated $V_\kappa I – V_\kappa IV$) and at least five λ sub-groups (designated ($V_\lambda I – V_\lambda V$). Similar analysis of human heavy chains has revealed that there are four subgroups of the V_H regions (designated $V_H I – V_H IV$). Detailed information on amino acid sequences of immunoglobulin is to be found in reference [10].

3.4 The antibody binding site

Since Porter demonstrated the antibody binding activity in the Fab fragments of immunoglobulins, a great deal of interest has been shown in the location and structure of the binding site.

Experiments to determine the ability of isolated chains of antibody to bind antigen have generally revealed that isolated heavy chains retain some activity and light chains little or none. However, using the sensitive technique of fluorescence enhancement, antigen binding by L-chains has been demonstrated [11]. Recombination of isolated heavy and light chains from an anti-DNP antibody leads to recovery of up to 50% of the original antigen binding activity of the whole antibody [12] suggesting that both heavy and light chains are involved in the binding of the antigen. The existence of hypervariable regions or CDR in both heavy and light chain variable regions suggests that contact with antigen occurs at such regions and that the binding site is formed by the interaction of both chains. One experimental approach which has been used to provide support for this hypothesis and to obtain further information on the nature of the binding site is the technique of affinity labelling [12]. Essentially this involves reaction of anti-hapten antibody with a radioactive, chemically modified hapten which, by virtue of an additional reactive group, is able to form covalent bonds with amino acids in or near the antibody binding site. There are three major groups of affinity labelling reagents:

(1) Diazonium reagents, e.g. *p*-phenylarsonic acid diazonium fluoborate
(2) Bromacetyl derivatives, e.g. α, *N*-bromoacetyl ε-*N*-DNP-L-lysine (BADL)
(3) Arylnitrene reagents, e.g. the 4-azido-2-nitrophenyl (NAP) group.

The principle of the method is shown in Fig. 3.7. The affinity label (X–R) first forms a reversible complex C with the antibody binding site A. Whilst in the site, it forms a covalent bond with a suitable amino acid residue Y forming the stable complex L (i.e. $A + X–R \rightleftharpoons C \rightarrow L$). Subsequent digestion and analysis of the molecule allows the labelled amino acid to be characterized. By such methods early results showed that the tyrosine residues in the variable regions of both heavy and light chains are labelled, with the ratio of labelling of heavy:light ranging from 3:1–4:1. Several affinity labelling reagents are now being actively used, and results seem to indicate that the hypervariable regions are indeed susceptible to labelling. Studies of this type should help to define the amino acids involved in the binding site. Homogeneous antibodies of defined specificity also provide another approach to the study of the binding site. However, the precise determination of the three dimensional structure of the antibody site

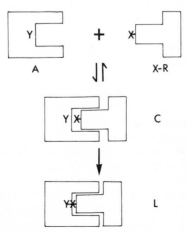

Fig. 3.7 The principle of affinity labelling [13].

requires the additional use of X-ray crystallographic techniques. The human myeloma protein NEW binds a γ-hydroxy derivative of vitamin K_1. Crystallized complexes of this antibody and the antigen have been analysed by X-ray crystallography [14] and the results have shown that binding of the antigen by the antibody consists of multiple points of contact with 10–12 amino acids in the hypervariable regions of both heavy and light chains (Fig. 3.8). The antigen thus nestles in a groove or cleft formed by the heavy and light chain variable regions.

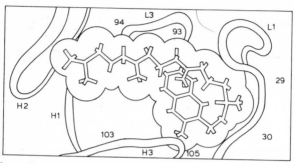

Fig. 3.8 The antibody combining site. The antigen molecule nestles in a cleft formed by the antibody combining site. The example shown is based on X-ray crystallography studies of human IgG (the myeloma protein NEW) binding γ-hydroxyl vitamin K. The antigen makes contact with 10–12 amino acids in the hypervariable regions of both heavy and light chains. The numerals refer to amino acids identified as actually making contact with the antigen. L1 and L3 refer to the hypervariable (CDR) of the light chain and H1, H2, H3 the hypervariable (CDR) of the heavy chain involved in binding the antigen (after [14]).

Work on the combination region of a mouse IgA myeloma 460, which binds competitively the haptens dinitrophenol and menadione, has suggested that binding cleft contains sites which can bind to both antigenic determinants. Furthermore, it has been shown that the two sites appear to be separated by a distance of 1.2–1.4 nm. These studies thus suggest that an individual combining region of an antibody may bind diverse determinants at different sites within the combining region [15]. The existence of polyfunctional binding sites would clearly have important implications for our understanding of antibody specificity and the genetic control of antibody diversity.

3.5 Immunoglobulin domains

Edelman *et al.* [15], in their determination of the complete amino acid sequence of a human IgG immunoglobulin showed that the variable region of the heavy chain extends from the N-terminal to approximately residue 120 and the constant region from residue 121 to 450. These workers noted that there were three regions within the constant region of the heavy chain which were very similar to each other and to the constant region of the light chains C_L. It has been suggested that these homology regions, C_H1, C_H2 and C_H3 of the heavy chain, and C_L of the light chain and the variable regions of H and L chains are folded into compact 'domains' [17]. Each of these domains is stabilized by an intrachain disulphide bond and, it is proposed, has one or more functions e.g. antigen binding for the V_L and V_H domains; complement binding sites located in the C_H2 domain and macrophage binding for the C_H3 domain. A function for the C_H1 domain has yet to be precisely defined. Fig. 3.9 shows an outline of the overall structure of the IgG protein sequenced by Edelman *et al.* [16] showing the variable and constant regions of both chains, homology regions, the disulphide bond arrangement and the position of carbohydrate.

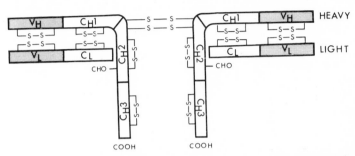

Fig. 3.9 A schematic representation of the general structure of human IgG1 myeloma protein Eu. V_H, V_L represent variable regions of the heavy and light chains, C_L, C_H1, C_H2, C_H3 represent the constant region domains of the light and heavy chains, respectively. CHO represents the carbohydrate (after [15]).

X-ray studies of crystallized myeloma proteins have revealed that the polypeptide chains of the domains are folded in a characteristic fashion. There is virtually no α helical structure in the immunoglobulins and the predominant feature is the presence of anti-parallel β-pleated sheets stabilized by intrachain disulphide and hydrogen bonds. The basic immunoglobulin 'fold' is shown in Fig. 3.10 and a three dimensional diagram representing the folding of V and C region domains is given in Fig. 3.11.

All the domains (with the exception of the C_H2 domain) are arranged in pairs which are held together by non-covalent forces and a schematic representation of this is shown in Fig. 3.12 for an IgG1 molecule. This diagram illustrates the lateral domain interactions which occur in immunoglobulins i.e. V_L–V_H, C_L–C_H1, C_H3–C_H3. Note that carbohydrate is bound to C_H2 (which does not have lateral interactions) and presumably stabilizes this domain. The possibility exists that longitudinal interactions occur, perhaps following binding with antigen, but the latter is a subject of much debate.

3.6 Allotypes and idiotypes of immunoglobulins

A number of structural variants of immunoglobulins exist, including the heavy chain determinants which are associated with the five immunoglobulin classes and the immunoglobulin sub-classes. These are all present in normal individuals and are termed isotypic variants and will be discussed in detail in Chapter 5. As discussed earlier, structural variations in the C_L

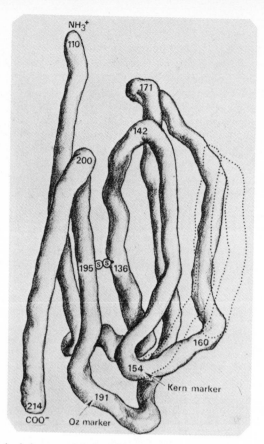

Fig. 3.10 The basic immunoglobulin fold. A model of the C-terminal half of a human λ light chain, in which the numbers represent the positions in the amino acid sequences. The constant region begins at residue 110. The dotted line indicates the position of the extra loop in the V_L region (from [18] with permission).

regions gives rise to the types and subtypes of light chains (κ, λ, oz$^+$, oz$^-$) and variations in the variable regions of both heavy and light chains are recognized as subgroups (V_κ I–IV and V_H I–IV). These variations in immunoglobulin structure are depicted schematically in Fig. 3.13 together with two further important sources of immunoglobulin variation – allotypes and idiotypes which have made important contributions to our understanding of the synthesis of antibody molecules.

3.6.1 *Immunoglobulin allotypes*

Allotypes or genetic variants of protein molecules are the products of allelic genes (alternative forms of a gene at a single locus) inherited in a simple Mendelian manner. They were first described in 1956 by Oudin [22] in rabbit immunoglobulins and by Grubb [23] in human immunoglobulins. Allotypes differ in individuals of the same species and correlate with differences in amino acid sequence of the protein; they are detected using

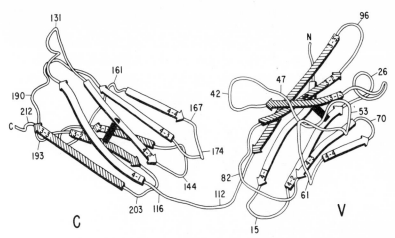

Fig. 3.11 The folding of V_L and C_L domains of a λ light chain. The white arrows (4-1—4-4) represent segments of the four-chain surface and the shaded arrows (3-1—3-3) represent segments of the three-chain surface — forming β-pleated sheets. Arrows point towards the C terminus of the chain and the black bars represent disulphide bridges (from [19] with permission).

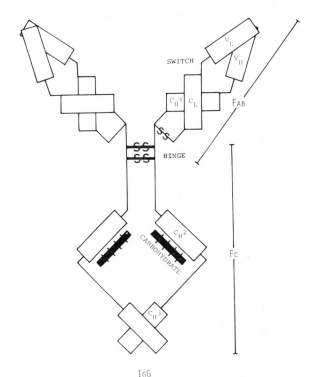

Fig. 3.12 A diagram of human IgG1 molecule showing the arrangement of domains from [20] with permission).

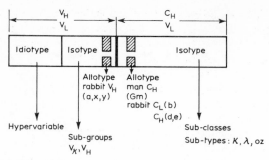

Fig. 3.13 Schematic representation of the variants of immunoglobulins (after [21]).

antibodies produced by immunizing an animal with immune complexes or immunoglobulins which have been obtained from another animal of the same species having immunoglobulin allotypes differing from those of the immunized animal. Several groups of allotypes (groups a, d, e, x and y) detected by immunological methods have been described for both the constant and variable regions of the H chains of rabbit IgG (Fig. 3.14). In addition, groups of allotypes in the C regions of IgA (groups f and g) and IgM (group n) have been described.

Allotypes also occur on the light chains (groups b and c) but it is the presence of such markers in both C_H and V_H of the heavy chain which make them useful for studies into the genetic control of antibody synthesis. A rabbit H chain may carry an allotypic specificity from each of groups a, d, and e. It has been demonstrated that in a rabbit which is heterozygous at both the light chain group (b group allotypes) and at the a group of the heavy chain, any one immunoglobulin molecule will carry only one of the two possible allelic markers of each group. Thus an animal which is geno-

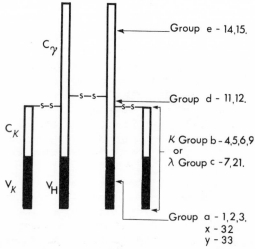

Fig. 3.14 Allotypes of rabbit IgG (after [24]).

typically a1, a3, b4, b6 will have immunoglobulin molecules which have the specificities a1, b4; a1, b6; a3, b4 and a3, b6. Both H and both L chains of each molecule have the same markers. This phenomenon is known as allotypic restriction. Of great interest was the surprising observation of Todd [25] that group a allotypes were present on both γ and μ chains and it has subsequently been shown that they are also present on α and ϵ chains (the 'Todd phenomenon'). More detailed information on rabbit allotypes is given in [26].

The existence of allotypes on human immunoglobulins was discovered fortuitously by Grubb. Sera from patients with rheumatoid arthritis contain rheumatoid factors (RF) which are antibodies (usually IgM but also other isotypes) that react with human IgG. Such RF thus agglutinate red cells coated with human anti-red blood cell antibodies. Grubb observed that RF from some patients only reacted with IgG from certain individuals and furthermore, any agglutination reaction could only be specifically inhibited by sera from individuals whose immunoglobulins carried the same antigenic determinants as expressed on the antibodies on the coated red blood cells. These antigens have been shown to be genetically determined and are designated Gm (genetic marker). Thus allotypes associated with the IgG1 subclass have the prefix G1m; those with IgG2, G2m and so on. The particular allotypes have been designated both alphabetically and numerically. The human allotypes, their location and the amino acid interchanges associated with each are shown in Table 3.1.

It has been shown that these markers are in the constant region of the heavy chain and are located in the C_H2 or C_H3 regions of the molecule with the exception of Gm_f and Gm_z which are in the C_H1 region. It is thought that the subclass specific antigens are related to amino acid sequence differences. For example, it appears that there are two amino acid differences between residues 355 and 358 in the Fc fragments of IgG Gm(a) and IgG1 Gm(a⁻) protein (Table 3.1).

A second system genetically closely linked to the Gm system has been described. This is the $Am2^+$ marker on heavy chains of IgA2 proteins. An independently inherited marker in the constant region of human κ light chains, the Inv (or Km) marker, constitutes the third allotypic marker system of human immunoglobulins. A single amino acid substitution at residue 191 differentiates the Inv1 from the Inv3 determinant.

Human λ light chains exist in two forms which differ by one amino acid substitution at residue 190. In the Oz(+) subtype, lysine is at residue 190 and in the second type, the Oz(−) chains, the amino acid is arginine. These two types are present in all normal humans making it unlikely that they are allelic forms of a single gene locus. The subject of human immunoglobulin genetics is reviewed in reference [27].

Allotypes have also been demonstrated in the C_H region (usually C_H2 or C_H3) of the heavy chains of mouse immunoglobulin classes (IgG1, G2a, G2b, M, A and E) but not in the V_H or light chains. Since there are many inbred strains of mice (individuals of an inbred strain are genetically identical) the study of mouse allotypes has been simplified compared to that of the human and rabbit. Anti-allotype antibodies can be raised by

Table 3.1 Examples of human immunoglobulin allotypes, their designation localization and amino acid interchanges related to them

Immunoglobulin class	Gm allotype		Location of determinant	Amino acid responsible for serological specificity
	Alphabetical designation	Numerical designation		
				355 356 357 358
IgG1	G1m(a)	G1m(1)	C_H3	−Arg−Asp−Glu−Leu
	G1m(non-a)	G1m(-1)	C_H3	−Arg−Glu−Glu−Met
IgG1	G1m(f)	G1m(3)	C_H1	214(Arg)
	G1m(z)	G1m(17)	C_H1	214(Lys)
IgG2	G2m(n)	G2m(23)	C_H2	
IgG3	G3m(b°)	G3m(11)	C_H3	436(Phe)
	G3m(non-b°)	G3m(−11)	C_H3	436(Tyr)
	G3m(g)	G3m(21)	C_H2	296(Tyr)
	G3m(non-g)	G3m(−21)	C_H2	296(Phe)
				308 309 310
IgG4	G4m(4a)		C_H2	−Val−Leu−His−
	G4m(4b)		C_H2	−Val———His−
IgA2		A2m(1)		
		A2m(2)		
Kappa light chains	Inv(1)		C_K	191(Leu)
	Inv(2)		C_K	191(Val)

immunizing mice with immunoglobulins from a donor bearing allotypes different to that of the recipient and using this approach it has been shown that each immunoglobulin class and subclass has its own set of allotypic determinants. The inbred strains of mice fall into ten groups which differ on the basis of the allotypic markers on the heavy chains of the IgG2a subclass [28].

3.6.2 Immunoglobulin idiotypes
As discussed above, antibodies can be obtained which recognize isotypic and allotypic determinants on immunoglobulin molecules. However it is also possible to obtain antibodies to individual immunoglobulin molecules or the products of single clones of antibody producing cells. The determinants involved are called idiotypes and were discovered in the following way. Antibodies to *Salmonella* from an individual rabbit A were injected into recipient rabbits with the same allotypes. Some of these recipients produced antibodies which reacted specifically with rabbit A anti-salmonella antibodies but not with other immunoglobulins from A nor with

anti-salmonella antibodies from other rabbits [29]. Similarly, a rabbit antibody to an individual human myeloma protein, when absorbed with other myelomas having the same isotype and allotypes as the immunizing protein, reacted specifically with the unique determinants of the immunogen [30]. These observations led to the concept of idiotypy. The antigenic determinants which generate antibody formation are known as idiotypic determinants (idiotopes) and the group of idiotopes on the immunizing antibody is called the idiotype. The antibodies to these are called anti-idiotope and anti-idiotypic antibodies, respectively. Idiotypic determinants are located on the variable regions of the antibody, as shown by:

(1) Testing fragments of the antibodies which consist of the variable regions
(2) Inhibition of the idiotype–anti-idiotype reaction by the specific ligand (hapten)
(3) Correlation of amino acid sequence data with idiotypy.

It is very likely that the hypervariable regions function as the idiotype. The observation that not all anti-idiotype–idiotype reactions can be blocked by the hapten suggests that the anti-idiotype reacts with hypervariable regions that are not involved in the binding site.

Certain idiotypes are shared by many antibodies induced by an antigen in animals of the same genetic background (shared or public idiotypes). These shared idiotypes are in some cases coded by genes in the germline. Each immunoglobulin molecule has a unique or private idiotype which is present only on molecules produced by the same clone.

The idiotypes of antibodies produced in response to a foreign antigen are not normally present on the immunoglobulins of the animal and are thus themselves potential immunogens. When the concentration of such antibodies reaches a critical level, they may thus induce auto-anti-idiotypic antibodies. It has been proposed that regulation of the immune response is in part achieved by the interactions of regulatory T-cells with specificity for such idiotypic determinants [31] through a network of idiotype–anti-idiotype interactions between T and B lymphocytes.

References
[1] Porter, R. R. (1959), *Biochem J.*, **73**, 119.
[2] Nisonoff, A., Wissler, F. C. and Lipman, L. N. (1960), *Science*, **132**, 1770.
[3] Edelman, G. M. and Poulite, M. D. (1961), *J. Exp. Med.*, **113**, 861.
[4] Porter, R. R. (1962), in *Basic Problems of Neoplastic Disease*, Columbia University Press, New York.
[5] Porter, R. R. (1967), The Structure of Antibodies, *Scientific American*, p. 81.
[6] Valentine, R. C. and Green, N. M. (1967), *J. Molec Biol.*, **27**, 615.
[7] Cathou, R. E. (1979), in *Immunoglobulins* (eds G. W. Litman and R. A. Good), Plenum, New York, p. 37.
[8] Putnam, R. W., Titani, K., Wikler, M. and Shinoda, T. (1967), *Cold Spring Harbor Symposium on Quantitative Biology*, **32**, 9.
[9] Wu, T. T. and Kabat, Ed. (1970), *J. Exp. Med.*, **132**, 211.

[10] Kabat, E. A., Wu, T. T. and Bilofsky, H. (1979), Sequences of Immunoglobulin Chains, NIH Bethesda, USA, Publication Number 80–2008.

[11] Yoo, T. J., Roholt, O. A. and Pressman, D. (1967), *Cold Spring Harbor Symposium on Quantitative Biology*, 32, 117.

[12] Haber, E. and Richards, F. (1967), *Proc. Roy. Soc.*, **166B**, 176.

[13] Singer, S. J. and Dolittle, R. F. (1964), *Science*, **153**, 13.

[14] Amzel, L. M., Polijak, R. J., Saul, F., Varga, J. M. and Richards, F. F. (1974), *Proc. Nat Acad Sci USA*, **71**, 1427.

[15] Richards, F. F., Varga, J. M., Rosenstein, R. W. and Konigsburg, W. H. (1978), in *Immunochemistry, an Advanced Textbook* (eds L. E. Glynn and M. W. Steward), John Wiley, Chichester, p. 59.

[16] Edelman, G. W., Cunningham, B. A., Gall, W. E., Gottleib, P. D., Rutishauser, H. and Waxdall, M. J. (1969), *Proc. Nat. Acad. Sci. USA*, **63**, 78.

[17] Edelman, G. W. and Gall, W. E. (1969), *Am. Rev. Biochem.*, **38**, 415.

[18] Poljak, R. J. (1975), *Nature (Lond)*, **256**, 373.

[19] Schiffler, M., Girling, R. L., Ely, K. R. and Edmunson, A. B. (1973), *Biochemistry*, **12**, 4620.

[20] Marquart, M. and Deisenhofer, J. (1982), *Immunology Today*, **3**, 160.

[21] Roitt, I. M. (1980), *Essential Immunology*, 4th edn, Blackwell Scientific, Oxford, p. 46.

[22] Oudin, J. (1956), *Compt. Rend.*, **242**, 2606.

[23] Grubb, R. (1956), *Acta Path. Microbiol. Scand.*, **39**, 195.

[24] Todd, C. W. (1972), *Fed. Proc.*, **31**, 188.

[25] Todd, C. W. (1963), *Biochem. Biophys. Res. Commun.*, **11**, 170.

[26] Kindt, T. J. (1975), *Advanc. Immunol.*, **21**, 35.

[27] Natvig, J. and Kunbel, H. G. (1973), *Advanc. Immunol.*, **16**, 1.

[28] Green, M. C. (1979), *Immunogenetics*, **8**, 89.

[29] Oudin, J. and Michel, M. (1963), *Compt. Rend. Acad. Sci. (Paris)*, **257**, 805.

[30] Stater, R. J., Ward, S. M. and Kunkel, H. G. (1959), *J. Exp. Med.*, **101**, 85.

[31] Jerne, N. J. (1974), *Annal Immunol. (Paris)*, **125C**, 373.

4   Antibody—antigen interaction

Perhaps the most important function of antibody molecules is to combine with the corresponding antigen to form an antibody—antigen complex which is then eliminated from the circulation by the cells of the mono-nuclear phagocytic system. The specific interaction of antibody with the antigen involves the interaction of the antigen binding site, (which is comprised of the hypervariable regions of both heavy and light chains), with the antigenic determinant.

4.1   The intermolecular forces involved in antibody—antigen interactions

The intermolecular forces which contribute to the stabilization of the antibody—antigen complex are the same as those involved in the stabilization of the configuration of proteins and other macromolecules. A brief account of these forces will be given here since they are fundamental to the specificity of the antibody—antigen interaction and a simple diagrammatic representation of them is given in Fig. 4.1. Detailed physico—chemical and mathematical consideration of these forces can be obtained from any physical chemistry text book.

4.1.1   *Hydrogen bonding*

Hydrogen bonding results from the interaction of an H atom covalently linked to an electronegative atom with the unshared electron pair of another electronegative atom. In antibody—antigen interactions, amino or

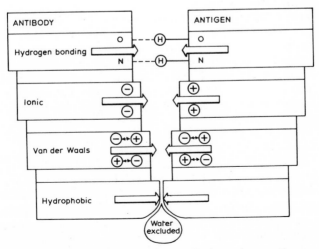

Fig. 4.1 The intermolecular forces involved in antibody—antigen reactions.

hydroxyl groups are the major hydrogen donors. The following are examples of H bonds:

$$-O \ldots H - O-$$

$$-O \ldots H - N-$$

$$-N \ldots H - N-$$

4.1.2 *Apolar or hydrophobic interaction*
Non-polar or hydrophobic side chains of amino acids such as leucine, isoleucine, valine or phenylalanine do not form hydrogen bonds with water. Thus in an aqueous environment, such groups prefer to interact with each other rather than with water − this is termed apolar or hydrophobic interaction. The stability of these bonds is due to structural altera- tion of the aqueous environment when the groups come together and exclude water. The proteins concerned thus take up a lower energy state and gain entropy when combined, resulting in a nett attractive force between them. These forces play a major part in antibody−antigen inter- actions.

4.1.3 *Ionic or coulombic interaction*
This interaction is the result of attraction between oppositely charged groups on two side chains i.e.

$$R-COO^- \ldots {}^+NH_3-R$$

where R represents the protein side chain. While it is clear that an inverse relationship exists between the charge on an immunogen and the charge of the antibody it elicits, these coulombic forces are not dominant or essential in the stabilization of antibody−antigen complexes [1].

4.1.4 *Van der Waal's forces*
As two polar groups come together their electron clouds interact. This interaction involves the induction of oscillating dipoles in the two mole- cules which results in an attractive force. If the electron clouds are com- plementary to each other they are able to get closer and the attractive force is greater. The Van der Waal's attractive forces vary inversely with the sixth power of the distance between the groups involved i.e. $F \propto I/d^6$. Similar attractive forces, called London dispersion forces, operate between non-polar groups.

4.1.5 *The steric factor or steric repulsive forces*
The attractive forces discussed above increase as the distance between the interacting groups decrease. Thus coulombic forces are inversely propor- tional to the second power of the distance and Van der Waal's forces to the sixth power. However, steric repulsion is much more sensitive and varies inversely with twelfth power of the distance. This repulsive force between non-bonded atoms arises from the interpenetration of their electron clouds. Thus the better the complementarity of the electron

cloud shapes the lower will be the repulsive force. This force is therefore the basis of antibody discrimination. A non-homologous antigenic determinant will not have an electron cloud which is complementary to that of the antibody binding site and thus the repulsive force will be high and attractive forces will be minimized. This may result in the antibody having low affinity for the particular determinant. Conversely, where the electron clouds of the antigenic determinant and antibody are complementary, that is, there is a 'good fit', then steric repulsion will be minimized and the attractive intermolecular forces maximized giving a high affinity antibody. Therefore, antibody affinity is the summation of these attractive and repulsive forces (see Fig. 4.2). The concept of antibody affinity, its measurement and biological significance are discussed later (see Section 4.4 and 4.6).

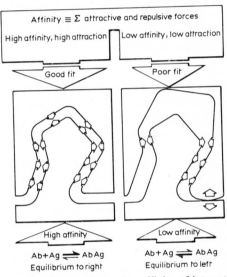

Fig. 4.2 The concept of antibody affinity. The affinity of interaction between antibody and antigen results from the summation of attractive and repulsive forces.

4.2 The measurement of antibody—antigen reactions

In general, tests for the measurement of antibody—antigen reactions make use of the fact that after combination with antigen, antibodies will agglutinate red cells or particles coated with antigen or potentiate measureable biological reactions or form precipitates which can be visualized in solution or gel. Some of these methods are shown in Table 4.1 together with the minimum amount of antibody which they can detect.

There is a wide variation in sensitivity of such tests depending upon which particular property of the antibody is being exploited (i.e. precipitation, agglutination, etc.). The reaction of an antibody (Ab) with an antigen or antigenic determinant (Ag) to produce a complex (Ab—Ag) can

Table 4.1 Sensitivity of antibody detection methods [2]

Method	Minimum amount antibody detectable ($\mu g\ ml^{-1}$)
Precipitation	
(i) ring test	20–30
(ii) optimal proportions	50
(iii) P-80 radioprecipitation	1.0
Gel diffusion	
(i) double diffusion tube	20–50
(ii) single diffusion tube	12–110
(iii) double diffusion plate	40
(iv) single diffusion plate	10
Immunoelectrophoresis	100
Radioimmunoelectrophoresis	0.02
Crossed electrophoresis	100
Agglutination	0.1–120
Passive haemagglutination	0.03
Complement fixation	0.1–1.0 g
Passive cutaneous anaphylaxis	0.02
Farr binding (($(NH_4)_2SO_4$ precipitation)	0.05–1.0
Immunosorbent	20
Radioimmunoassay	μg–ng ml^{-1} range
Antiglobulin precipitation	μg–ng ml^{-1} range
Enzyme-linked immunosorbent assays	μg–ng ml^{-1} range

be represented as follows:

$$Ab + Ag \underset{k_d}{\overset{k_a}{\rightleftharpoons}} AbAg$$

where k_a is the association constant, k_d the dissociation constant. This is the primary antibody–antigen reaction and several methods are available to detect and quantify this interaction. After one of these primary interactions have occurred other secondary or tertiary manifestations [3] may or may not occur depending on the nature of the antibody being tested. Examples of secondary manifestations are precipitate formation, agglutination and complement fixation. Tertiary manifestations such as anaphylaxis or immune elimination of antigen are very far removed from the primary reaction and greatly affected, for example, by host variations. Since many variables control both secondary and tertiary manifestations of the primary antibody–antigen reaction, negative results with these tests occur even in the presence of antibody demonstrable by the primary tests. Therefore, since the results of primary binding tests are the only reliable guide to the presence or absence of antibody, experiments requiring accurate estimations of antibody levels should include at least one such test. Such methods of antibody detection and quantification are classified into primary, secondary and tertiary categories in Table 4.2.

Table 4.2 Classification of antibody detection methods [3]

Category	Method
Primary	Radioimmunoelectrophoresis
	Farr binding
	Antiglobulin technique
	Fluorescence quenching, enhancement, polarization
	Equilibrium dialysis
Secondary	Gel diffusion
	'P-80' radioprecipitation
	Agglutination
	Complement fixation
Tertiary	Passive cutaneous anaphylaxis
	Arthus reaction
	Immune elimination

4.3 The study of the chemistry of antibody—antigen reactions

In order to study the antibody—antigen interaction in detail it is generally necessary to determine in some way the free and bound forms of either antigen or antibody after they have interacted.

The simplest method for partitioning insoluble antibody—antigen complexes from soluble reactants is the precipitation reaction (a secondary test). For many years this was the only test available to immunochemists and detailed studies of this process have provided much information of the nature of antibody—antigen interactions. Arrhenius (1907) was the first to apply the Law of Mass Action to immunochemical reactions in his attempts to describe mathematically the immune precipitation reaction. We now know that this law does not apply to the complex secondary reactions involved in precipitate formation but this work is nevertheless, a landmark in the history of immunochemistry.

4.3.1 Precipitation reactions

The work of Arrhenius and those who followed him was seriously hampered by the fact that although immune precipitates could be measured, their actual composition could not be chemically quantified since both antibody and antigen were proteins. For similar reasons, the determination of antibody and antigens in the supernatant was not possible. This problem was overcome by Heidelberger and Kendall [5] by using pneumococcal polysaccharides as antigens which do not interfere with antibody nitrogen determinations. More recently, of course, the application of radioisotope techniques has eliminated these difficulties. However, using polysaccharide antigens, Heidelberger and his colleagues [6] have described the generalized form of the immunological precipitation reaction which is shown in Fig. 4.3.

As increasing amounts of antigen are added to a fixed amount of antibody, the quantity of antibody precipitated increases. After the addition

41

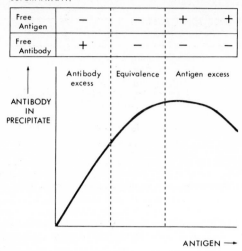

Fig. 4.3 The quantitative precipitin reaction.

of small amounts of antigen, some precipitate is formed and free antibody is detectable in the supernatant – this is the antibody excess zone. Here the ratio of antibody to antigen in the immune complex depends on the valency of the antigen. The addition of larger amounts of antigen results in increased precipitation until a point is reached where no free antibody or antigen is detectable in the supernatant. This is the equivalence zone. Marrack [7] hypothesized that at this point, optimal proportions of antibody and antigen form a continuous, stable antibody–antigen 'lattice' which precipitates. At high levels of antigen, free antigen appears in the supernatant and at the same time, precipitate formation is still maximal. This is the first stage of the antigen excess zone. At conditions of extreme antigen excess, the amount of precipitate is markedly reduced, due to the formation of soluble complexes. Solubilization results from the excess free antigen competing for the antibody sites in the precipitate with subsequent formation of soluble complexes with the molar composition Ab_1Ag_2 or Ab_2Ag_3 (see Fig. 4.4).

Both the quantity and quality (e.g. affinity, immunoglobulin class, etc.) of antibody is important in determining whether a precipitate will be formed. Antigenic valency is also a critical factor since the formation of a lattice is impossible if the antigen is monovalent. Recent evidence however, suggests that antibody–antigen precipitate formation does not depend solely on antigen-binding site interaction but also upon interactions between the Fc region of the antibody molecules involved [8].

4.3.2 *Agglutination reactions*
The formation of a lattice with antibody and a multivalent antigen results in precipitation as discussed above. However, when antibody reacts with antigenic sites on the surface of particles such as those on red blood cells

42

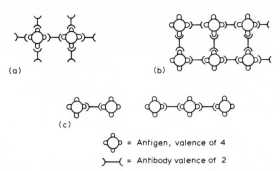

Fig. 4.4 A diagrammatic representation of the antibody—antigen complexes formed in (a) in antibody excess; (b) at equivalence and (c) in antigen excess. The antibody is shown as a bivalent molecule, the antigen as a tetravalent molecule.

or bacteria, agglutination occurs. Inert particles such as latex, coated with antigen are also agglutinated by antibody (passive agglutination). Essentially what occurs is that sufficient antibody—antigen bonds are formed to overcome the natural repulsive charge effects of cells, which results in their aggregation or agglutination. As with precipitation reactions, the valency of the antigen and the nature of the antibody are important in agglutination reactions. Cells with few antigenic determinants will be less readily agglutinated than those with many determinants. The multivalent IgM antibody is, on a molecule to molecule basis, a more efficient agglutinating antibody than the divalent IgG. Univalent antibodies or incomplete antibodies (antibodies which do not agglutinate or precipitate antigen for physico—chemical or structural reasons) can be demonstrated by the Coombs test. In this agglutination test, the antibodies react with, but do not agglutinate cells (or cell coated with antigen). Agglutination is produced by the subsequent addition of antibodies to the incomplete antibody, made in another species. These three types of agglutination reactions are shown in Fig. 4.5.

4.4 The thermodynamics of the antibody—antigen reaction

As described in Section 4.2 the events which may or may not follow the primary interaction of antibody with the homologous antigen have been classified as secondary or tertiary manifestations. In view of the complexity of variables which influence these secondary and tertiary manifestations, studies attempting to investigate the detailed nature of antibody—antigen reactions have, in general, utilized primary binding techniques, some of which are listed in Table 4.3.

4.4.1 *Antibody affinity*

The term antibody affinity refers to the strength of the interaction between an antigenic determinant and the homologous antibody binding site. In effect it is the summation of the attractive and repulsive forces described in Section 4.1. Thus a high affinity antibody is one which forms a strong bond with the antigenic determinant to give an antibody antigen complex with a low tendency to dissociate. On the other hand, a low affinity anti-

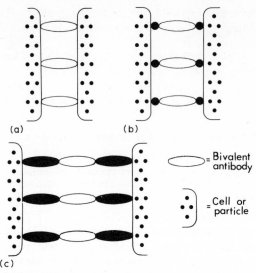

Fig. 4.5 Agglutination reactions: (a) active agglutination; (b) passive agglutination with antigen, coated onto cell or particle; (c) Coombs test: agglutination by incomplete antibodies (shaded) only on addition of anti-γ-globulin (open).

body forms a complex with antigen which requires less energy for dissociation. Thus the higher the affinity of the antibody, the greater will be the amount of antigen bound to antibody at equilibrium. Affinity is a thermodynamic measurement of the strength of the antibody–antigen interaction and is expressed either as the equilibrium constant K (with units of litres mole^{-1}) or as the standard free energy change ΔG° (with units of kilo calories mole^{-1}). The quantitative relationship of the interaction between antibody and antigen at equilibrium is represented by the equation:

$$Ab + Ag \underset{k_{\mathrm{d}}}{\overset{k_{\mathrm{a}}}{\rightleftharpoons}} AbAg$$

Table 4.3 Quantitative primary binding tests

1 Equilibrium dialysis
2 Fluorescence quenching
3 Fluorescence enhancement
4 Fluorescence polarization
5 Separation of complexes from free antigen by
   (i) ammonium sulphate precipitation
   (ii) antiglobulin precipitation
   (iii) ultracentrifugation
   (iv) gel filtration
   (v) polyacrylamide disc electrophoresis
   (vi) adsorption of antigen onto charcoal, silica
6 Temperature jump method
7 Stopped flow methods

where *Ab* represents free antibody, *Ag* free antigen, *AbAg*, the antibody—antigen complex, k_a and k_d the association and dissociation constants, respectively.

The Law of Mass Action states that the rate of formation of complex is proportional to the concentration of the reactants. Thus the rate of association is equal to $k_a[Ab][Ag]$, and the rate of dissociation is equal to $k_d[AbAg]$. At equilibrium the rates of association and dissociation are equal, thus:

$$k_a[Ab][Ag] = k_d[AbAg]$$

$$\therefore \quad \frac{k_a}{k_d} = K = \frac{[AbAg]}{[Ab][Ag]} \tag{4.1}$$

where K is the equilibrium constant. Also $\Delta G^\circ = -RT\ln K$ where R is the gas constant and T the absolute temperature. Thus a high affinity antibody will have a more negative ΔG° value compared to that of a low affinity antibody.

At this point, a note of clarification is necessary. In the literature, the terms 'affinity' and 'avidity' are often used synonymously. Affinity is a thermodynamic expression of the primary binding energy of antibody for an antigenic determinant. Experimentally this term has its most precise application in monovalent hapten—anti-hapten systems. Avidity on the other hand, although it depends in part on affinity, also involves factors such as antibody valency, antigen valency and other nonspecific factors associated with binding. For example, a multivalent IgM antibody whose binding sites have the same affinity for antigen as a bivalent IgG antibody, will have a greater avidity for a multivalent antigen than the IgG antibody. The IgM is a more avid antibody because its multiple binding sites give it a greater ability to bind the antigen. The avidity of an antibody is often expressed in terms of its ability to effect a biological antigen binding function, such as virus neutralization. This clearly involves factors other than primary antibody—antigen binding. Thus affinity and avidity are not synonymous and have been termed 'intrinsic' and 'functional affinity', respectively [9].

4.4.2 *Derivation of equations for affinity calculation*

Thermodynamic measurement of antibody affinity requires that the reactants, antigen and antibody, are pure and in solution. Ideally the reactants should be homogeneous with regard to antigenic determinants and binding sites. These requirements are in fact met when monoclonal anti-hapten antibodies are used which, in many instances, behave in an ideal manner [10]. Such measurements with polyclonal antibodies are complicated because of their heterogeneity and multivalency for antigen. However, in spite of these limitations, reasonably precise affinity measurements can be made with antibodies to a monovalent hapten. Affinity measurements with antibody—antigen systems which are less thermodynamically precise can be made but such determinations only provide relative affinity values, e.g. with anti-protein antibodies.

The equilibrium constant K, or affinity, of isolated specific anti-hapten antibody can be determined experimentally by measurement of free hapten and antibody bound hapten at equilibrium over a range of hapten concentrations by a variety of methods (see Table 4.3 and Section 4.4.3 below). The following is a brief outline of the derivation of equations frequently used for affinity calculations using the data from such experiments. Detailed derivations can be obtained from references [11] and [12].

From the Law of Mass Action the following form of the Langmuir adsorption isotherm may be derived:

$$\frac{[AbAg]}{[Ab]} = r = \frac{nK[Ag]}{1 + K[Ag]} \tag{4.2}$$

where r = moles hapten bound per mole of antibody present, $[AbAg]$ = bound antibody concentration, (Ab) free antibody concentration, $[Ag]$ free hapten concentration, n = antibody valence, K = equilibrium constant.

Thus a plot of $r/[Ag]$ versus r (Scatchard plot) for a range of antigen concentrations allows n, the antibody valence and K to be obtained (Fig. 4.6).:

$$\frac{r}{[Ag]} = nK - rK \tag{4.3}$$

This equation is frequently used to obtain values for the intrinsic association constant K_0 in systems involving a divalent antibody ($n = 2$) and a monovalent hapten. When half the antibody sites are bound (i.e. $r = 1$) then (4.3) becomes

$$\frac{1}{[Ag]} = 2K - K = K_0$$

Thus K_0 is equal to the reciprocal of the free hapten concentration at equilibrium when half the antibody sites are bound to hapten.

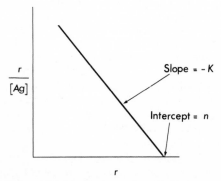

Fig. 4.6 Scatchard plot of ideal antibody–antigen binding.

An alternative affinity equation can be obtained from either (4.2) or (4.3):

$$\frac{1}{r} = \frac{1}{n} \frac{1}{Ag} \frac{1}{K} + \frac{1}{n}$$

(4.4)

A plot of $1/r$ versus $1/Ag$ (Langmuir plot) allows both antibody valence (n) and affinity K to be derived (Fig. 4.7). Both the Scatchard and Langmuir equations should give rise to straight line plots. However, with most antibodies, even when isolated and purified, there is deviation from linearity. The reasons for this deviation are complex but it is basically due to the heterogeneity of antibody — not all antibody molecules in an antibody population have the same affinity for the hapten. This deviation makes affinity determination using these equations difficult. Various mathematical techniques have been used to overcome this problem. The distribution of antibodies of various affinities within an antibody population is assumed to be describable in terms of a Gaussian or Sipsian function. By using the logarithmic transformation of the Sips equation [1, 13].

$$\frac{r}{n} = \frac{(K_0 [Ag])^a}{1 + (K_0 [Ag])^a}$$

(4.5)

The following equation is obtained:

$$\log \frac{r}{n - r} = a \log K_0 + a \log[Ag]$$

[4.6]

where a = heterogeneity index. Therefore, in a plot of $\log r/n - r$ versus $\log [Ag]$ where

$$\log \frac{r}{n - r} = 0, \text{ then } K_0 = \frac{1}{[Ag]}$$

K_0 determined in this way is thus the peak of a presumed normal distribution curve. The heterogeneity index, a, is given by the slope of the line (Fig. 4.8).

Thus in this way, the average intrinsic association constant K_0 and the

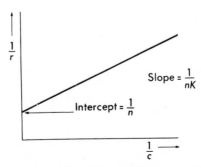

Fig. 4.7 Langmuir plot of ideal antibody–antigen binding.

47

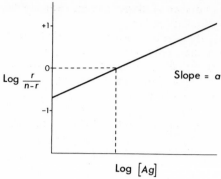

Fig. 4.8 Sips plot of antibody–antigen binding.

heterogeneity index can be calculated. As the heterogeneity index approaches 1 the antibody population approaches homogeneity with regard to association constants.

In all that has been described so far, it is assumed that the antibody is pure and the amount present is known accurately. These calculations also require a knowledge of the valence of the antibody. In situations where isolation and purification of antibody is not possible or not desirable, calculation of K can be made with respect to total antibody binding sites (Ab_t) rather than to antibody concentration and valence [14]. Thus (4.2) becomes:

$$\frac{b}{[Ab_t]} = \frac{K[Ag]}{1 + K[Ag]}$$

from which

$$\frac{1}{b} = \frac{1}{Ab_t} \frac{1}{[Ag]} \frac{1}{K} + \frac{1}{Ab_t} \tag{4.7}$$

Where b = antibody-bound antigen and $[Ag]$ free antigen. Thus measurement of bound and free antigen over a range of antigen concentrations and plotting the data according to (4.7) makes the determination of Ab_t (total antibody binding sites) possible; i.e. when $1/[Ag] = 0$, then $1/b = 1/Ab_t$.

Affinity and heterogeneity index can be determined by substituting b for r and Ab_t for n in (4.6) [15].

The heterogeneity of affinity within an antibody population makes it necessary for the measurement of the equilibrium concentrations of bound and free antigen (or antibody) to be made over a range of free antigen (or antibody) concentrations. Measurements of average affinity have assumed that affinity heterogeneity could be described by a normal Gaussian distribution function or Sips function. Estimates of heterogeneity indices have been obtained from binding data in hapten–anti-hapten systems and it was assumed that serum antibody was the sum of the antibodies synthesized by a large number of clones of cells each producing antibody of a different affinity for antigen [16].

The possibility that the heterogeneity of affinity of serum antibody was not describable in these terms was suggested by Pressman *et al.* [17] and it now appears that affinity heterogeneity is not distributed in a Gaussian manner but can be skewed or even bimodal [18]. At the present time, it is not possible adequately to estimate affinity distribution based on experimental data from binding studies. These observations have far-reaching implications concerning our ability to measure 'average' affinity and affinity heterogeneity and no completely satisfactory resolution of the problem is currently available.

4.4.3 *Methods of affinity measurement*

Basically, the measurement of antibody affinity depends upon the determination of free and antibody-bound antigen at equilibrium. Because of antibody heterogeneity such determinations are usually carried over a range of antigen concentrations to ensure that all antibody binding sites are saturated. This requires the separation of bound and free antigen either by a dialysis membrane, selective precipitation by salt or by antibody, gel filtration or by techniques which do not dissociate the bound antigen. Other methods utilize the alteration of some property of the antibody or antigen such as changes in fluorescence properties. A survey of the principal methods which have been used for the determination of antibody affinity is given in Table 4.4. The most commonly used techniques will be briefly described here.

(a) Equilibrium dialysis

This technique is widely accepted as the standard method of affinity measurement [20]. A dialysis membrane is used to partition antibody-bound hapten and free hapten. Antibody solution (purified antibody or immunoglobulin fraction of serum) is placed on one side and radioactive hapten on the other side of a dialysis membrane which allows the hapten to diffuse through it, but not the antibody. With time, hapten diffuses into the antibody compartment and some binds to the antibody. Equilibrium is eventually achieved when the free hapten concentration is the same on both sides of the membrane (Fig. 4.9). The radioactivity inside the antibody compartment represents bound and free hapten and that outside represents free hapten. Thus by subtraction, the amount of bound hapten is obtained. This process is repeated for several hapten concentrations using the same antibody concentration. The affinity of the antibody may then be calculated from the data by the methods described above.

(b) Ammonium sulphate precipitation

In this technique, the insolubility of antibody and hence antibody–hapten complexes in 50% saturated ammonium sulphate is used [21]. Bound and free radioactive hapten concentrations can be readily determined by counting the precipitates and supernatants after precipitation with ammonium sulphate. This technique has been used for affinity measurement

Table 4.4 Commonly-used methods for affinity measurement [17]

Method	Principal applicable antigens	Affinity parameter measured
Equilibrium dialysis	Haptens, dialyzable molecules	Intrinsic*

Changes in fluorescence properties
    Quenching
    Enhancement
    Polarization
	Haptens, antigens with specific fluorescence properties	Intrinsic and functional†

Precipitation of complexes by
    Ammonium sulphate
    Polyethylene glycol
    Antiglobulins
    Ethanol ammonium acetate
	All antigens soluble in precipitants	Intrinsic and functional

Electrophoretic separation of complexes	Haptens, proteins, peptides	Intrinsic and functional

Size separation
    Ultracentrifugation
    Gel chromatography
    Equilibrium molecular sieving
	Haptens, proteins, peptides	Intrinsic and functional

Phage neutralization	Haptens	Functional
Equilibrium filtration	Viruses	Functional
Antibody plaque-forming cell inhibition	Soluble antigens	Functional

* Intrinsic affinity ≡ affinity
† Functional affinity ≡ avidity

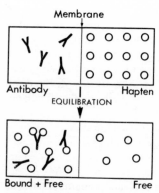

Fig. 4.9 Diagram illustrating the principle of equilibrium dialysis.

[19, 22] and the results obtained compare reasonably well with those determined by equilibrium dialysis [10, 22]. It is ideal for rapid estimations of affinity and has the advantage of not requiring purified antibody. Whole serum can be used as a source of antibody. This method has also been applied to the measurement of the relative affinity of antibodies to protein antigens [19, 23]. However, in systems other than those involving hapten antigens, application is limited to those antigens which are soluble in 50% sulphate.

(c) Fluorescence quenching

A molecule fluoresces when it absorbs light of one wavelength and then dissipates the absorbed energy by the emission of light at a longer wavelength. Proteins fluoresce when irradiated with ultraviolet light. Although phenylalanine, tyrosine and tryptophan amino acid residues are all potentially fluorescent, the tryptophan residues make the major contribution to this fluorescence. Purified antibodies irradiated by light of wavelength between 280–295 nm emit light of wavelength 330–350 nm due to the fluorescence of their tryptophan residues. If this excitation energy is transferred to a non-fluorescent molecule the protein fluorescence is decreased. Thus when a purified antibody reacts with a hapten having certain fluorescence properties, the excitation energy produced on irradiation with UV light is transferred to the bound non-fluorescent hapten and is dissipated by non-fluorescent processes resulting in decreased or quenched antibody fluorescence. The method of fluorescence quenching was first described by Velick *et al.* [24] using antibodies to the 2,4 dinitrophenyl (DNP) group. The hapten DNP-lysine absorbs light maximally at 360 nm, and its absorption spectrum overlaps the antibody emission spectrum (Fig. 4.10) and is thus particularly suited to the study of antibody

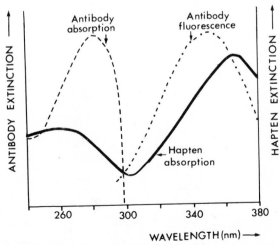

Fig. 4.10 Absorption and emission spectra of rabbit anti-DNP antibody and DNP-lysine hapten (after [25]).

51

fluorescence quenching. The technique is quite straightforward. The maximum quenching (Q max) which can be obtained with all antibody sites occupied by hapten is first determined (up to 80% is possible). Then, assuming a linear relationship between quenching and number of antibody sites bound, the number of antibody sites bound at any given hapten concentration can be readily determined, and the antibody affinity calculated.

The method has the advantage of requiring small amounts of antibody but is limited to highly purified antibodies to haptens with the necessary spectral properties.

(d) Fluorescence enhancement

Changes in hapten fluorescence as a result of binding to antibody have been utilized for antibody affinity measurement. With certain fluorescent haptens, combination with antibody results in diminution of protein fluorescence but instead of the transferred excitation energy from the protein typtophan being dissipated, (as in the case with non-fluorescent haptens) the fluorescent hapten absorbs the energy and shows an increased fluorescence. This is known as fluorescence enhancement, and this property of certain haptens has been utilized for antibody affinity determinations. It has the distinct advantage of not requiring purified antibody since the fluorescence properties of the hapten and not those of the antibody are being measured.

An example of a molecule with such fluorescence properties is the dimethylaminonaphthalene-sulphonamide (DANS) group [24, 25]. This has an absorption maximum where tryptophan fluorescence is maximum and an absorption minimum where protein absorbs maximally. Maximum fluorescence is emitted at 520 nm where proteins do not fluoresce. Thus when rabbit antibodies react with DANS-lysine, the fluorescence of the ligand is raised by 25–30 fold [25, 26] (see Fig. 4.11). This increase in

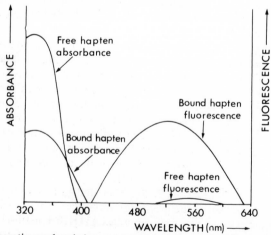

Fig. 4.11 Absorption and emission spectra of free and antibody-bound DANS-lysine (after [25]).

fluorescence is related to the number of hapten molecules bound and can therefore be used to quantify bound and free hapten at equilibrium for affinity determination. It is particularly useful for measuring antibodies of low affinity. For example, it has been used for demonstrating hapten binding by free light chains which are of very low affinity.

(e) Fluorescence polarization

The fluorescent emmission from a small molecule is not normally polarized because the molecules are randomly orientated during the short time interval between excitation and emission. The amount of rotation of a molecule as a result of rotary Brownian motion decreases as molecular size increases. Thus, when a fluorescent molecule reacts with an antibody molecule, the size is obviously increased and the rotatary movement is restricted. The process of random orientation of the molecules in this situation is slower than for the free fluorescent molecule and this results in the polarization of the fluorescent emission. The extent of fluorescence polarization can be used to quantify bound and free antigen and facilitate affinity determinations [27]. This method is applicable to the study of haptens and fluorescent labelled protein antigens and their interaction with the corresponding antibodies. The common labelling reagents for protein antigen studies are fluorescein isothiocyanate and dimethylamino-naphthalene sulphonyl chloride. The method is subject to certain problems particularly with larger antigens and is appraised by Parker in reference [25].

(f) Other methods

Several other methods for affinity determination are available including the use of the ultracentrifuge to quantify the amount of bound and free antigen. The ultracentrifuge has been used to measure the affinity of rheumatoid factors for human IgG [28]. Rheumatoid factors are present in the serum of patients with rheumatoid arthritis, and are antibodies to immunoglobulins ('anti-globulins') which react with Fc antigenic determinants on the patients own IgG.

The use of gel filtration to provide antigen binding data for antibody affinity calculations has been described [29]. In this method, Sephadex beads (cross-linked dextran) of carefully chosen sizes are used to partition antibody-bound antigen and free antigen on the basis of their size. Larger molecules (e.g. the antibody–antigen complex) are excluded from the beads whereas smaller molecules (i.e. free antigen) are able to enter the beads. This method has been applied to the measurement of the affinity of rheumatoid factors for isolated fragments of human IgG. These fragments pFc' bear the antigenic determinants to which the rheumatoid factor antibody activity is directed [30].

All the methods discussed so far have applied to serum antibodies. A technique for measuring the avidity of antibodies at the level of the antibody forming cell has also been described [31]. The method involves the determination of the amount of free antigen (protein of hapten) which will inhibit plaque-forming cells in the Jerne technique [32]. High affinity antibody-producing cells require less free antigen to inhibit haemolytic plaque formation than do low affinity antibody-producing cells.

4.5 The kinetics of the antibody–antigen reaction

If the antibody–antigen reaction is represented as a reversible reaction:

$$Ab + Ag \underset{k_{21}}{\overset{k_{12}}{\rightleftharpoons}} AbAg$$

where Ab represents free antibody sites; Ag, free antigen; $AbAg$, the antibody–antigen complex and k_{12} and k_{21} the association and dissociation rate constants, respectively, then the rate of formation of antibody–antigen complex is given by:

$$\frac{d[AbAg]}{dt} = k_{12}[Ab][Ag] - k_{21}[AbAg]$$

and at equilibrium the net rate is zero, then

$$\frac{k_{12}}{k_{21}} = \frac{[AbAg]}{[Ag][Ag]} = K, \text{ the equilibrium constant.}$$

It should thus be possible to study the kinetics of this reaction by conventional techniques involving addition of reactants, mixing and the subsequent determination of the concentration of the reactants or products at time intervals. Unfortunately, this simple type of approach is not suitable for the study of the reaction between antibody and complex antigens. The interaction of polyvalent antibodies with polyvalent antigen results in the formation of complexes and aggregates at rates which do not necessarily depend upon the primary interaction of antibody and antigen. These problems of aggregation can be minimized by the use of very low concentration of reactants or by performing the determination in antigen excess [33] but the problem of the heterogeneity of determinants on complex polyvalent antigens remains an obstacle to meaningful interpretation of data obtained.

The simple hapten–antihapten reaction, coupled with more sophisticated techniques and instrumentation, has been used to investigate the kinetics of the antibody–antigen reaction. Even so, these reactions also pose problems for the investigator. Kinetic methods require precise measurements of changes in the concentration of reactants and products as a function of time. However, it has been known for many years that antibody–antigen reactions occur quickly but it has been established that hapten–antihapten interactions are among the fastest of known biological reactions. This has posed problems for their kinetic measurement: conventional chemical kinetic methods cannot be used because in these the time of mixing is the rate-limiting step and equilibration occurs before changes in concentration of reactants and products can be determined. However, the use of high dilution of reactants in the range of $10^{-8}-10^{-10}$ M^{-1} overcomes this problem since at these concentrations mixing is not the rate-limiting step and the production of radioactively labelled haptens with very high specific activity has enabled dilutions of antigen in this range to be conveniently employed [34]. Rapid mixing of reactants has also been achieved by the use of the stopped-flow fluorimetric technique [35] and has allowed kinetic measurements of the hapten—antihapten reaction. The kinetics of fast reactions have been studied by the temperature-jump relaxation technique which was first applied to the investigation of antibody-hapten kinetics by Froese, Sehon and Eigen [36]. In addition, ammonium sulphate globulin precipitation has been used to investigate antibody—antigen kinetics [33] and, more recently, the kinetics of antihapten—hapten interactions have been studied by utilizing dextran-coated charcoal to separate the free hapten from bound hapten [37].

Work of several authors using both the stopped-flow and temperature jump relaxation techniques has shown that the values for k_{12} for all hapten—antihapten systems tested was within one order of magnitude of 10^8 M^{-1}. On the other hand, marked variation in the dissociation rate constants, k_{21}, was observed. Froese [38] demonstrated that a tenfold difference in average association constant was due to differences in k_{21} rather than to variations in k_{12} suggesting that the stability of the hapten—antihapten antibody complex is governed by the dissociation rate constant. This observation, plus those of other workers have clearly demonstrated that it is the dissociation rate constant k_{21} which determines the affinity of antibody for hapten.

4.6 The biological aspects of antibody affinity

4.6.1 *Factors affecting antibody affinity*
A range of factors are known to influence antibody affinity and these include the nature of the immunogenic stimulus, genetic factors, qualitative and quantitative aspects of lymphocyte function, dietary and hormonal factors, reticuloendothelial function and the effects of free antibody and antigen—antibody complexes. The mechanisms by which these factors affect affinity are not clear, particularly since the cellular basis of the control of affinity has not been fully characterized. However, it does appear that the control of affinity is likely to be expressed at the level of the macrophage [39] or at the level of T-cell helper and suppressor functions [10, 41]. Obviously, the nature and mode of presentation of antigen will be important, particularly since a wide range of materials behave as immunological adjuvants and can raise antibody affinity values above those achieved by immunization without adjuvants [42] and may override any, perhaps subtle controls exerted by T-cells and macrophages on affinity.

The genetic control of antibody affinity has been demonstrated in mice and rats and is controlled by mechanisms independent of those governing antibody levels [43, 44]. Two lines of mice, one producing high affinity antibody, the other low affinity antibody, have been obtained by selective breeding based on the affinity of antibody induced to protein antigens injected in saline [45]. Since the two lines diverged gradually in their affinity for antigens during the selective breeding, antibody affinity is likely to be under the control of several genes.

The influence of dietary factors on affinity has been studied in animals and the results clearly demonstrate that protein restriction strikingly affects the affinity of antibody produced whilst antibody levels are not consistently affected [46]. This observation is of importance when evaluating the effects of malnutrition on antibody responses in man, where the effects on antibody levels have not been consistent. Immuno-depression induced by drugs [47] or infection [48] may also reduce antibody affinity and the excessive production of low affinity antibody in these ways may have important immunopathological significance (see below).

4.6.2 *The biological importance of antibody affinity*
Fundamental to the cell selection theory of Burnet is the suggestion that potential antibody-forming cells have immunoglobulin receptors on their surface with specificity for antigen. Cells which will ultimately make high affinity antibody will have high affinity receptors and low affinity producing cells will have low affinity receptors. Antigen 'selects' cells which are capable of making antibody to it and stimulates them to proliferate and secrete antibody. The heterogeneity of antibody affinities thus produced reflects the heterogeneity of the affinities of the lymphocyte receptors.

A wide range of anti-DNP antibody affinities has been demonstrated in the serum of an individual rabbit [16]. Ten antibody fractions were obtained by fractional precipitation from this serum which had affinities for DNP-lysine differing by up to 10 000-fold. This was the first experimental demonstration of the heterogeneity of antibody binding affinity produced by an individual animal. As discussed earlier, it has been assumed that this heterogeneity is continuous and can be represented by a Gaussian or Sipsian distribution function. However, it is possible that in fact limited heterogeneity [17] occurs in which only a restricted number of antigen-reactive cells are stimulated to proliferate and produce antibodies. Detailed investigations of antihapten–antibody binding affinities utilizing computational procedures to produce histograms describing the distribution of affinities in an antiserum, have shown that the distribution is highly complex and non-symmetrical [18]. In spite of such theoretical problems the measurement of antibody affinity under a variety of experimental situations has provided data which are consistent with predictions made on the basis of the cell selection hypothesis.

It has been known for some time that the avidity of antibodies e.g. anti-diphtheria toxin and also the affinity of antihapten antibodies increases ('matures') with time after immunization. These observations, together with the demonstration that both the level of antibody affinity and the rate of maturation of affinity with time are markedly reduced when large doses of antigen are used for immunization are consistent with the selection theory. In the presence of large amounts of antigen, cells of a wide range of affinities will be stimulated. This situation occurs early during immunization or after injection of large doses of antigen. As antigen becomes more limiting, competition for antigen by cells occurs, and those with high affinity receptors are more likely to be stimulated. Thus with time, antigen selects those cells which have the highest affinity receptors resulting in a progressive increase in the affinity of serum antibody. The validity of this explanation has been confirmed by measurement of the relative affinities of anti-DNP antibodies produced by lymph node cells in tissue culture taken from immunized rabbits at various times after immunization [49]. The affinity of the antibody synthesized in culture increased as the time interval between immunization and lymph node removal increased. Thus affinity maturation is due to changes in the population of cells producing antibody rather than to secondary effects in the serum after synthesis.

It is clear that the affinity with which cell receptors bind the homologous antigenic determinant is of importance in determining whether the cell is triggered to differentiate or to become non-responsive (or tolerant). Affinity alone does not appear to be the only governing factor in tolerance induction but there does seem to be a relationship between the functional affinity of the receptors and the ease with which susceptible cells are tolerized. Using a mathematical approach to the question of the binding of antigen by cell receptors, it has been suggested that the redistribution and aggregation of receptors induced after interaction with antigen may facilitate more stable binding of multivalent antigen to the cell. On mathematical grounds, it was argued that for multivalent antigens such as T-independent antigens, it is receptor density not receptor affinity which determines the number of antigen molecules bound per cell. Thus the processes of patch formation and capping, which increase the local density of receptors will result in a greater binding of antigen in affinity-independent manner. However, the affinity of the receptors (in addition to antigen valence and concentration) is important in determining the rate of patching and hence in the triggering of the cell [50].

The valence of IgM antibody is very much influenced by the size of the antigen used [51]. The observed IgM anti-dextran valence varies from 10 for dextran of M_r 342, to 5 for dextrans of M_r 7 000–237 000. For dextrans of M_r 1.87 x 10^6 the observed valence is 2.3. These results clearly indicate that steric hindrance plays a vital role in determining the valence of IgM for an antigen.

The maturation phenomenon has been described for several antigens, usually after immunization in adjuvant. However, when human serum albumin is injected without adjuvants, maturation occurs but subsequently the affinity of serum antibody falls. This has been demonstrated in both rabbits and mice. Furthermore, low affinity antibody populations have been shown to persist throughout the antibody response and in other studies, affinity has been shown to oscillate between high and low during the response. These observations are difficult to reconcile with the cell selection hypothesis which cannot easily take into account of these features of the antibody response. However, it is possible that in the absence of adjuvant (which in the case of Freund's complete adjuvant provides an antigen depot with continuous slow release of antigen over long periods), the decrease in affinity may be due to gradual death of short-lived high affinity antibody producing cells. The possibility must

57

also be considered that this fall may arise from the generation of suppressor T-cells perhaps operating within the idiotype—anti-idiotype network.

Since antibodies are multivalent, the question has been asked as to what is the functional significance of such multivalence. It has been known for some time that hapten-conjugated bacteriophage can be neutralized by anti-hapten antibody. Work on the neutralization of DNP-conjugated bacteriophage $\Phi \times 174$ by anti-DNP antibodies [9] has provided convincing evidence that there is a greater strength of interaction between a multivalent antibody and a polyvalent antigen, than between a multivalent antibody and a monovalent antigen. With the DNP-bacteriophage—anti-DNP system, the equilibrium constant of each antibody binding site for DNP-lysine (intrinsic affinity) can be determined by equilibrium dialysis. Also kinetic studies of the neutralization of bacteriophages bearing multiple DNP groups by anti-DNP allow calculation of equilibrium constants (functional affinity) in a polyvalent system. IgG anti-DNP antibodies (divalent) have an intrinsic affinity for DNP-lysine of 10^4 litres mole^{-1} and a functional affinity of 10^7 litres mole^{-1}, an enhancement due to divalency of 10^3-fold. With IgM anti-DNP antibodies, intrinsic affinity was 10^4-10^5 litres mole^{-1} whereas functional affinity was greater than 10^{11} litres mole^{-1} which represents an enhancement due to multivalency of 10^6-fold. These authors suggested that this represents multivalent IgM attachment to the conjugated phage via at least three of its binding sites. Similar work on DNP-T$_4$ bacteriophage [52] has provided confirmation of the concept that the neutralizing power of an antibody depends not only on concentration of antibody but also is a function of affinity, valence and molecular conformation. This enhancement of affinity due to multivalency represents a considerable advantage to the immune system in that considerably lower levels of multivalent antibody are required to provide an effective immune response than would be necessary with monovalent antibody.

As discussed above, there have been certain observations which suggest that affinity maturation does not always occur. However, if the formation of a progressively higher affinity antibody is a general phenomenon of the antibody response then it follows that high affinity antibody should be biologically more effective than antibody of lower affinity. A number of studies have confirmed this contention (Table 4.5). For example, the passive transfer of equal amounts of either high or low affinity antibody into recipient mice results in very marked differences in the ability of the recipient mice to eliminate subsequently injected ^{125}I-labelled antigen (Fig. 4.12). High affinity antibody achieves a greater elimination of the antigen than does low affinity antibody. Further examples of the superiority of high affinity antibody over low affinity antibody are shown in Table 4.6. High affinity antibody is far more efficient at complement fixation and passive cutaneous anaphylaxis reactions than in antibody of a 10-fold lower affinity. The persistence of low affinity antibody over a long period and the demonstration of high and low affinity memory cells undermines the early assumption that all memory cells have high affinity receptors. However, the function of this persisting low affinity antibody is not

Table 4.5 Biological reactions in which high affinity antibody is more effective than low affinity antibody

Neutralization of toxins
Haemagglutination
Haemolysis
Complement fixation
Passive cutaneous anaphylaxis
Immune elimination of antigen
Membrane camage
Virus neutralization
Protective capacity vs. bacterial infections
Enzyme inactivation
High affinity *Ab:Ag* complexes localize in mesangium of kidney
Low affinity *Ab:Ag* complexes localize in basement membrane of kidney

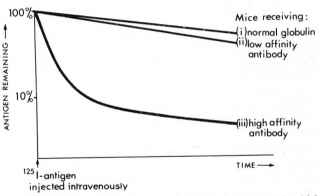

Fig. 4.12 The elimination of antigen by mice which had received either high or low affinity antibody [53].

Table 4.6 The effect of antibody affinity on the activity of antibodies [54]

	Antibody preparation	
	A	B
Affinity, litres mole^{-1}	1×10^7	1.1×10^6
Antibody required for +ve PCA (μg ml^{-1})	31.3–52.5	>500
Concentration for complement fixation (2 units)		
Antibody (μg ml^{-1})	2.10	8.30
Antigen (μg ml^{-1})	0.03	0.08

59

completely clear, but since high affinity cells are more readily tolerized than are low affinity cells, the presence of the low affinity cells will be of clear advantage to the host by providing first-line defence against the antigen.

4.6.3 *The immunopathological significance of antibody affinity*

An excessive production of low affinity antibody has been considered as an expression of immunodeficiency. According to this view, which is based on work in rabbits and in mice, low affinity antibody plays an important part in determining susceptibility to chronic antigen—antibody complex disease. Based on the clinical observations that immunodeficiency is broadly antigen non-specific, is likely to be under genetic control and that individuals can produce functionally poor antibodies, it was proposed that antigen—antibody complex disease arises as a result of a genetically controlled low affinity antibody response which fails to eliminate antigen and favours the production and subsequent tissue localization of antigen-excess complexes. The complexes deposited in tissues may therefore contain either low affinity antibody, or perhaps that small proportion of high affinity antibody present in a population on average of low affinity [43].

To test this hypothesis, two lines of mice, one producing high and the second low affinity antibody to protein antigen, were generated by a selective breeding programme [45]. After daily injections of serum albumin (which has been shown to induce a chronic disease in rabbits) chronic antigen—antibody complex disease was more severe in mice selected to produce a lower average affinity antibody response than those selected to produce higher affinity antibody. The low affinity mice had:

(1) Higher levels of circulating complexes
(2) More intensely staining deposits of antigen—antibody complexes and complement in the glomeruli of the kidneys
(3) Glomerular basement membrane deposition of complexes
(4) A more severe impairment of glomerular filtration [55, 56].

Similar results have been described after prolonged immunization of rabbits with ovalbumin, where development of diffuse membranous glomerulonephritis was associated with the production of non-precipitating antibodies of low functional affinity and mesangioproliferative glomerulonephritis with precipitating antibodies of high functional affinity. In passive acute nephritis in mice, differences were observed in the renal localization of passively transferred DNP-protein—anti-DNP complexes composed of rabbit antibodies with differing affinities. Complexes with high affinity anti-DNP antibodies localized predominantly in the mesangium whereas complexes with lower affinity antibody localized on the basement membrane. In a similar model it has been shown that passively transferred complexes made with rabbit antibodies of high functional affinity prepared over a wide range of antigen excess localized in the mesangium of mouse kidneys whereas complexes prepared with low functional affinity antibodies gave rise to diffuse proliferative glomerlonephritis with subepithelial deposits. The detailed references to this work can be obtained from reference [57].

These studies indicate that antibody affinity is likely to be an important factor in the production and subsequent localization of injurious antigen–antibody complexes. The genetically controlled production of low affinity antibody is thus associated with the development of experimental chronic disease but the mechanisms involved are not at present clear. The immuno-chemical characteristics of localized complexes, once defined, will help to illuminate the precise role played by antibody affinity in chronic antigen–antibody complex disease.

4.7 The specificity and cross-reactivity of antibody–antigen interactions

Antibody–antigen interactions can show a high degree of specificity. This means that the antigen combining sites of antibodies directed towards determinants on one particular antigen are not complementary to deter-minants on a second antigen. This specificity means that antibodies to a virus like influenza will react with and confer immunity to influenza virus but will not react with an unrelated virus such as poliomyelitis virus. The basis of this specificity derives from the reactions of a population of antibody molecules within an antiserum. Each antibody molecule within the antiserum to antigen A reacts with a different part of the antigen molecule and perhaps with different parts of the same antigenic determin-ant, (e.g. determinants X, Y and Z), and the combined reactivity of all the antibody molecules in the serum gives the antiserum its specificity (Fig. 4.13). However, when some of the determinants on antigen A are shared by a second antigen B (e.g. determinant Y) (Fig. 4.13(b)) then a proportion

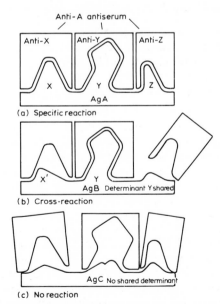

Fig. 4.13 The principle of antibody specificity and cross reactivity. (a) Antiserum to antigen A (anti-XYZ) reacts specifically with antigen A; (b) anti-A cross reacts with antigen B through recognizing determinant Y and partial recognition of X; (c) anti-A shows no reaction with antigen C which has no shared determinants.

of the antibody molecules directed to antigen A will also react with antigen B. Furthermore if the antiserum to A also weakly recognizes another determinant on B (e.g. X^1) then this will also contribute to this cross-reactivity of antiserum to X. Obviously, if no antigenic determinants are shared between antigen A and antigen C, there will be *no reaction* of anti-A with antigen C (Fig. 4.13(c)).

These two properties of antibody—antigen interactions are clearly illustrated by work performed many years ago by Landsteiner. He demonstrated that when the three isomers of amino benzene sulphonic acid (*ortho, meta* and *para*) were diazotized and coupled to a protein carrier and injected into rabbits, the antibodies obtained could differentiate between the three isomers. The anti-*meta*-aminobenzene sulphonate antiserum could clearly distinguish between haptens in which the tetrahedral sulphonate group is in the *meta, ortho* or *para* position by reacting more strongly with the *meta* substituted hapten (Table 4.7). The antiserum does not react at all with closely related aminobenzene arsonates substituted in either the *ortho* or *para* positions. Furthermore, cross-reactivity is demonstrated in this system since the anti-*m*-aminobenzene sulphonate antiserum also reacts with the *ortho* and *para* substituted sulphonate and with *meta*-aminocarboxylate haptens.

Thus, antibody raised to a sulphonate group also reacts, though weakly, with the arsonate group. This common reactivity is explained by the fact that the two groups both possess negative charge and a tetrahedral structure, but because the arsonate group is larger and possesses an extra hydrogen atom, it reacts less well with the anti-sulphonate antibody. The same antibody gives an even weaker reaction with the carboxylate group which, though negatively charged, has a planar configuration. These findings indicate that antibody specificity is more a function of the overall configuration of the antigen than its chemical composition. It thus appears

Table 4.7 Specificity and cross-reactivity of antibodies illustrated by the reaction of anti-*meta*-amino benzene sulphonate antibodies with different isomers of amino benzene sulphonate and with the three isomers of the related amino benzene arsonate and amino benzene carboxylate antigens.

Nature of group R	ortho-	meta-	para-
Sulphonate (tetrahedral)	+ +[‡]	+ + +[*]	±[‡]
Arsonate (tetrahedral)	—	+[‡]	—
Carboxylate (planar)	—	±[‡]	—

[*] Specific reaction   [‡] Cross-reaction

likely that antibodies are directed against particular electron cloud shapes rather than towards particular chemical structures.

Landsteiner also demonstrated similar specificity of the antibody response to carbohydrate antigens. He synthesized the two isomeric diazo-tartranilic acids (D- and L-) and coupled them to a carrier protein. Antibodies to these conjugates would distinguish between these D- and L-isomers.

The ability of antibodies to distinguish between small differences in the nature of an antigen is further illustrated by the fact that an antiserum to *para*-aminophenyl β glucoside can discriminate between *para*-aminophenyl β glucoside and *para*-aminophenyl galactoside, a difference which involves just the interchange of a hydrogen and an hydroxyl on one carbon atom. Furthermore, antibodies to *para*-aminophenol α glucoside can distinguish between the α and the β glucosides (Table 4.8). Finally, the exquisite specificity of antibodies has also been shown by studies on antibodies to the loop peptide of hen egg white lysozyme [5, 8].

The isolated peptide containing the sequence 64–83 of lysozyme has been called the 'loop' peptide since it has the disulphide bridge between residues 64–80. When conjugated to multi poly-DL-ala-poly-lysine the resulting conjugate elicited antibodies to native lysozyme in rabbits and goats. Antibodies of similar specificity could be isolated from serum of animals immunized with lysozyme using a preparation of insolubilized 'loop' peptide ('loop' peptide coupled to bromacetyl cellulose). These isolated antibodies still reacted with lysozyme and this reaction could be inhibited by the 'loop' peptide but not by the open chain peptide derived from the 'loop' by reduction and carboxymethylation. This indicates that the antibodies to the 'loop' are recognizing a specific steric conformation rather than just an amino acid sequence. Similar results have also been shown with antibodies raised to a completely synthetic 'loop' peptide of virtually identical amino acid sequence.

Table 4.8 Specificity and cross-reactivity of antibodies illustrated by the reaction of anti-*para*-amino phenol α glucoside, β glucoside and β galactoside antibodies with the α and β glucosides and the β galactoside antigens.

	Antigen		
	p-aminophenol-α-glucoside	p-aminophenol-β-glucoside	p-aminophenol-β-galactoside
Antibody	Precipitation reaction		
Anti-α glucoside	+ + +*	+ +†	−
Anti-β glucoside	+ +†	+ + +*	−
Anti-β galactoside	−	−	+ + +*

* Specific reaction
† Cross reaction

Table 4.9 The affinity of the binding of various haptens by antibody raised to the 2,4 dinitrophenyl-L-lysyl group of a DNP-protein conjugate (data from [16])

$$
\begin{array}{c}
\text{CO} \\
| \\
\text{HC}-\text{CH}_2-\text{NH}-\underset{\text{NO}_2}{\overset{\text{NO}_2}{\bigcirc}} \\
| \\
\text{HN} \\
|
\end{array}
$$

Hapten	Affinity litre^{-1} mole^{-1} x 10^5
$\begin{array}{c} \text{COOH} \\ \| \\ \text{HC}-(\text{CH}_2)_4-\text{NH}-\bigcirc\text{NO}_2 \\ \| \\ \text{NH}_2 \end{array}$ (NO$_2$) ε-DNP-L-lysine	200
$\begin{array}{c} \text{COOH} \\ \| \\ \text{HC}-(\text{CH}_2)_3-\text{NH}-\bigcirc\text{NO}_2 \\ \| \\ \text{NH}_2 \end{array}$ (NO$_2$) δ-DNP-L-ornithine	80
$\text{H}_2\text{N}-\bigcirc\text{NO}_2$ (NO$_2$) 2,4-dinitroaniline	20
$\bigcirc\text{NO}_2$ (NO$_2$) m-dinitrobenzene	8
$\text{H}_2\text{N}-\bigcirc\text{NO}_2$ p-nitroaniline	0·5

The data shown in Table 4.9 illustrate the ability of an antibody raised against the 2,4 dinitrophenyl group to discriminate between various related haptens. This discrimination is illustrated by the fact that the affinity of the antibody for the hapten increases as the test hapten approximates the structure of the haptenic group on the immunogen. The immunogen was a 2,4 dinitrophenyl protein conjugate in which the hapten is coupled to the ε-amino group of lysine — the antibody bound ε-DNP-lysine with high affinity. The affinity of the antibody for the other haptens depended on their structural relatedness to ε-DNP-lysine [16].

Up to this point, antibody specificity has been described as resulting from the properties of the antibody molecules within the antiserum. In recent years, it has been suggested that any one antibody combining site may be complementary to a number of different antigenic determinants. The binding of these antigenic determinants is competitive and data suggest that there are spatially separated binding sites within the antigen combining site. For example, human myeloma protein 460 binds both menadione and dinitrophenol competitively and it seems that there are two distinct binding sites 1.2–1.4 nm apart within the binding cleft of this antibody which bind the two antigens [59]. Thus antibody molecules may be polyfunctional. According to this view, the specificity of a population of antibodies would not necessarily result from all the antibodies possessing the same specificity. If a large number of different polyfunctional antibodies all had a site which could react with a particular determinant, the nett reactivity of these antibodies would be high to that determinant but low to all other antigenic determinants. Thus antibody specificity would be viewed as an average characteristic of all the antibody molecules in the serum.

References

[1] Karush, F. (1962), *Advan. Immunol.*, **2**, 1.

[2] Gill, T. J. III (1970), *Immunochemistry*, **7**, 99.

[3] Farr, R. S. and Minden, P. (1968), *Ann. NY Acad. Sci.*, **154**, 107.

[4] Minden, P., Anthony, B. F. and Farr, R. S. (1968), *J. Immunol.*, **102**, 832.

[5] Heidelberger, M. and Kendall, F. E. (1929), *J. Exp. Med.*, **50**, 809.

[6] Heidelberger, M. (1939), *Bact. Rev.*, **3**, 49.

[7] Marrack, J. R. (1938), *The Chemistry of Antigen and Antibodies*, HMSO, London.

[8] Moller, N. P. H. and Steensgaard, J. (1979), *Immunol.*, **38**, 641.

[9] Hornick, C. L. and Karush, F. (1972), *Immunochemistry*, **9**, 325.

[10] Stanley, C., Lew, A. M. and Steward, M. W. (1983), *J. Immunol. Meths.* (in press).

[11] Day, E. D. (1972), *Advanced Immunochemistry*, Williams and Wilkins, Baltimore, p. 181.

[12] Steward, M. W. and Steensgaard, J. (1983), *Antibody Affinity: Thermodynamic Aspects and Biological Significance*, CRC Press, Boca Raton, Florida.

[13] Sips, R. (1948), *J. Chem. Phys.*, **16**, 490.

[14] Nisonoff, A. and Pressman, D. (1956), *J. Immunol.*, **80**, 417.

[15] Steward, M. W. and Petty, R. E. (1972), *Immunology*, **22**, 747.

[16] Eisen, H. N. and Siskind, G. W. (1964), *Biochemistry*, **3**, 966.

[17] Pressman, D., Roholt, D. A. and Grossberg, A. L. (1970), *Ann. NY Acad. Sci.*, **169**, 65.

[18] Werblin, T. and Siskind, G. W. (1972), *Immunochem.*, **9**, 987.

[19] Steward, M. W. (1978), in *Handbook of Experimental Immunology* (ed. D. M. Weir), 3rd edn, Blackwell, Oxford, p. 161.

[20] Eisen, H. N. (1964), *Methods in Medical Research*, **10**, 106.

[21] Farr, R. S. (1958), *J. Infect. Dis.*, **103**, 239.

[22] Stupp, V., Joshida, T. and Paul, W. E. (1969), *J. Immunol.*, **103**, 624.

[23] Steward, M. W. and Petty, R. E. (1972), *Immunol.*, **23**, 881.

[24] Velick, S. F., Parker, C. W. and Eisen, H. N. (1960), *Proc. Nat. Acad. Sci. (Wash.)*, **46**, 1470.

[25] Parker, C. W. (1973), in *Handbook of Experimental Immunology* (ed. D. M. Weir), 2nd edn, Vol. 1, Blackwell, Oxford, p. 14.

[26] Parker, C. W., Yoo, T. J., Johnson, M. C. and Colt, S. M. (1967), *Biochemistry*, **6**, 3408.

[27] Dandliker, W., Schapiro, H. C., Meduski, J. W., Alonso, R., Feigen, G. A. and Hamrick, J. R. Jnr. (1964), *Immunochem.*, **1**, 165.

[28] Normansell, D. E. (1970), *Immunochemistry*, **7**, 787.

[29] Stone, M. J. and Metzger, H. (1969), *J. Biol. Chem.*, **243**, 5049.

[30] Steward, M. W., Turner, M. W., Natvig, J. B. and Gaarder, P. I. (1973), *Clin. Exp. Immunol.*, **15**, 145.

[31] Anderson, B. (1970), *J. Exp. Med.*, **132**, 77.

[32] Jerne, N. K. and Nordin, A. A. (1963), *Science*, **140**, 405.

[33] Talmage, D. (1960), *J. Infect. Dis.*, **107**, 115.

[34] Smith, T. W. and Skubitz, K. M. (1975), *Biochemistry*, **14**, 496.

[35] Chance, B. (1963), in *Techniques in Organic Chemistry Part 3* (ed. A. Weissberged), Vol. 8, Interscience, New York.

[36] Froese, A., Sehon, A. H. and Eigen, M. (1962), *J. Chem.*, **40**, 178.

[37] Skubitz, K. M. and Smith, T. W. (1975), *J. Immunol.*, **114**, 369.

[38] Froese, A. (1968), *Immunochem.*, **5**, 253.

[39] Morgan, A. G. and Soothill, J. F. (1975), *Nature (Lond.)*, **254**, 711.

[40] De Kruyff, R. and Siskind, G. W. (1979), *Cell Immunol.*, **47**, 134.

[41] De Kruyff, R. and Siskind, G. W. (1980), *Cell Immunol.*, **49**, 90.

[42] Petty, R. E. and Steward, M. W. (1977), *Immunol.*, **32**, 49.

[43] Soothill, J. F. and Steward, M. W. (1971), *Clin. Exp. Immunol.*, **9**, 193.

[44] Steward, M. W. and Petty, R. E. (1976), *Immunology*, **30**, 789.

[45] Katz, F. E. and Steward, M. W. (1975), *Immunology*, **29**, 543.

[46] Reinhardt, M. C. and Steward, M. W. (1979), *Immunology*, **38**, 735.

[47] Steward, M. W., Alpers, J. H. and Soothill, J. F. (1973), *Proc. Roy. Soc. Med.*, **66**, 808.

[48] Pattison, J., Steward, M. W. and Targett, G. A. T. (1983), *Clin. Exp. Immunol.*, **53**, 175.

[49] Steiner, L. A. and Eisen, H. N. (1967), *J. Exp. Med.*, **126**, 1161.

[50] Perelson, A. S. (1978), in *Theoretical Immunology*, Vol. 8, Marcel Dekker, New York, p. 171.

[51] Edberg, S. C., Bronson, P. M. and Van Oss, C. J. (1972), *Immunochemistry*, **9**, 273.

[52] Blank, S. E., Leslie, G. A. and Clem, L. W. (1972), *J. Immunol.*, **108**, 665.

[53] Alpers, J. H., Steward, M. W. and Soothill, J. F. (1972), *Clin. Exp. Immunol.*, **12**, 121.

[54] Fauci, A. G., Frank, M. M. and Johnson, J. J. (1970), *J. Immunol.*, **105**, 215.

[55] Steward, M. W. (1979), *Clin. Exp. Immunol.*, **38**, 414.

[56] Devey, M. E. and Steward, M. W. (1980), *Immunology*, **41**, 303.

[57] Steward, M. W. and Devey, M. E. (1982), in *Immunological Aspects of Rheumatology* (ed. W. C. Dick), Medical and Technical Press, Lancaster, p. 63.

[58] Arnon, R. and Sela, M. (1969), *Proc. Nat. Acad. Sci. (USA)*, **62**, 163.

[59] Richards, F. F., Varga, J. M., Rosenstein, R. W. and Konigsberg, W. H. L. (1977), in *Immunochemistry, An Advanced Textbook* (eds L. E. Glynn and M. W. Steward), John Wiley, Chichester, p. 59.

5 Structure and biological activities of the immunoglobulin classes

Immunoglobulins in man exist in five classes: IgG, IgM, IgA, IgD and IgE, all of which contain the basic four-chain unit (Chapter 3). The heavy chains are designated γ, μ, α, δ and ϵ, respectively. The classes have different C_H regions but the V_H regions and the light chains are drawn from the same repertoire. In addition, human IgG exists in four subclasses and IgA, in two subclasses. The class and subclass of a particular immunoglobulin molecule is determined by the primary amino acid structure of the C_H regions, and in the case of the subclasses, the amino acid sequence differences are small. The immunoglobulin classes also differ in other respects — IgM is a pentameric molecule (containing five of the four-chain units plus an additional polypeptide, the J chain). Serum IgA can exist as monomers dimers or trimers and in secretions, secretory IgA is a dimer containing J chain and an extra component, the secretory component. These structural differences are related to the biological function which each class of immunoglobulin performs.

The physico-chemical characteristics of the human immunoglobulins are shown in Table 5.1 from which it can be seen that the immunoglobulins vary in molecular weight (and hence in their sedimentation coefficients in the ultracentrifuge). These differences arise from variations in the heavy chains. IgM and IgE heavy chains have an extra domain compared to the other classes and this leads to M_r of 72 000 and 73 000, respectively. The importance of the carbohydrate content of immunoglobulins has been discussed earlier but it can be seen from Table 5.1 that the proportion of carbohydrate varies from class to class and can be as high as 14%.

The serum concentrations and metabolic characteristics of the immuno-globulin classes in man differ markedly (Table 5.2). IgG is the predominant class and accounts for approximately 70% of the total immunoglobulin in human serum and is distributed almost equally between the intra- and extravascular compartments. IgM on the other hand, is found predominantly in the intravascular compartment. The metabolic characteristics (synthesis, catabolism) of each immunoglobulin class is different and the serum concentration of a particular protein is the result of the balance between synthesis and catabolism. Thus although the synthetic rates for IgG and IgA are similar (33 mg kg^{-1} day^{-1} and 24 mg kg^{-1} day^{-1}, respectively), the serum concentration of IgA is much lower than that of IgG because of the higher catabolic rate of IgA. The catabolic rate of IgG is proportional to the serum concentration whereas synthesis depends particularly upon antigenic stimulation. Thus in individuals with IgG myeloma proteins, there is an accelerated catabolic rate which results in the IgG having a very short half-life. In people with *hypo*gammaglobulin-aemia, the catabolic rate of IgG is reduced. Germ-free animals have low IgG levels but on leaving the germ-free environment the levels rise rapidly.

Table 5.1 Physico-chemical characteristics of human immunoglobulins [1]

Immunoglobulin	$M_r \times 10^{-3}$	Sedimentation coefficient (S)	H chain Designation	H chain $M_r \times 10^{-3}$	Number of domains	Carbohydrate (%)	L chain K/λ ratio
IgG1	146	7	γ_1	51	4	2–3	2.4
IgG2	146	7	γ_2	51	4	2–3	1.1
IgG3	165	7	γ_3	60	4	2–3	1.4
IgG4	146	7	γ_4	51	4	2–3	8.0
IgM*	970	19	μ	72	5	9–12	3.2
IgA1	160	7	α_1	58	4	7–11	1.4
IgA2	160	7(9,11,13)‡	α_2	58	4	7–11	1.6
sIgA†	405	11	α_1 or α_2	58	4	7–11	?
IgD	170	7	δ	63	4	10–14	0.3
IgE	188	8	ϵ	73	5	12	?

* IgM and sIgA have J chain
† sIgA has secretory component plus J chain
‡ Serum IgA also exists in polymeric forms, with J chain

Table 5.2 Serum concentration and metabolic characteristics of human immunoglobulins

Immunoglobulin	Normal serum concentration (mgs ml^{-1})	Intravascular distribution (%)	Half-life (days)	Fractional catabolic rate*	Synthetic rate (mg kg^{-1} day^{-1})
IgG1	9	45	21	7	33
IgG2	3		20	7	
IgG3	1		7	17	
IgG4	0.5		21	7	
IgM	1.2	80	5	9	3.3
IgA1	2.0	42	6	25	24
IgA2	0.5		6		
sIgA	0.05	–	–	–	–
IgD	0.03	75	3	37	0.4
IgE	0.0002	50	3	89	0.02

* Percentage of intravascular pool catabolized per day

For IgD and IgE there is an inverse relationship between serum concentration and catabolic rate, but note that these immunoglobulins are particularly important cell-bound molecules. The catabolic rate for both IgM and IgA is however, independent of their serum concentrations. In the remainder of this Chapter, the structure and properties of each of the human immunoglobulin classes will be discussed.

5.1 Immunoglobulin G

IgG is quantitatively the most important serum immunoglobulin and its major function is to neutralize toxins, viruses and to bind to and opsonize bacteria (i.e. enhance their phagocytosis and elimination). IgG is the only immunoglobulin to cross the human placenta and is thus a major defence mechanism against infection in the early part of the infant's life. There are four antigenically distinct subclasses of IgG: IgG1 is the major subclass (67% of the total IgG) followed by IgG2 (22%), IgG3 (7%) and IgG4 (4%). Fig. 5.1 represents the four-chain structure of IgG1 and shows also the positions of the inter- and intra-chain disulphide bonds, the site of cleavage by papain and the fragments thus produced and the location of the effector functions. The subclasses are antigenically cross-reactive but have subclass-

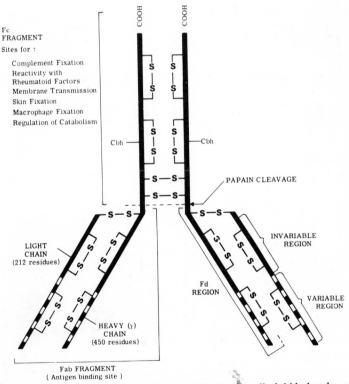

Fig. 5.1 The four-chain structure of human IgG1 showing disulphide bonds, papain fragments and the location of effector functions. Cbh represents site of attachment of carbohydrate (from [1] with permission).

specific amino acid sequences in the C_H regions. One particularly striking difference between the subclasses is the position and number of interchain disulphide bridges (Fig. 5.2). The number of bonds joining the heavy chains varies from 2 for IgG1 and IgG4, 4 for IgG2 and 10–13 for IgG3. The large number of disulphide bonds in IgG3 is associated with a long hinge region consisting of 62 amino acids (almost four times longer than for the other classes) and this results in the higher molecular weight of IgG3 compared to the other subclasses. The extended hinge region may also account for the efficiency of this subclass at activating the classical complement pathway (via Clq binding to the C_H2 region). Conversely, the failure of IgG4 to activate complement by this pathway may arise from steric hindrance preventing Clq binding to the $C_{\gamma_4}2$ region (see Fig. 5.2). Furthermore, there are marked differences in susceptibility of the subclasses to the enzyme papain (IgG3 is the most susceptible) which may also relate to the structure of the hinge regions.

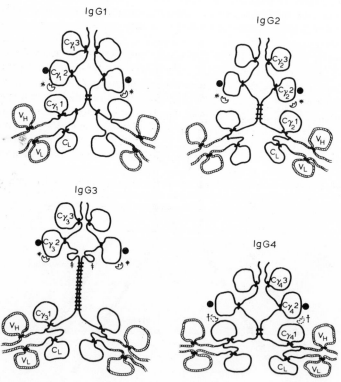

Fig. 5.2 The subclasses of human IgG showing the positions of the various domains (V_H, V_L, C_L and $C_{\gamma_1}1$, 2, 3, $C_{\gamma_2}1$, etc.), the position of the interchain disulphide bonds (●), the attachment site of carbohydrate (•), a speculative intrachain loop (‡) and the site for Clq binding (*); Clq binding to IgG4 can only occur in the isolated Fc fragment (†) (after [2] with permission).

5.2 Immunoglobulin A

IgA accounts for approximately 15–20% of the total immunoglobulin in human serum where it occurs as monomeric four-chain units (80%) and as polymeric forms (20%). The polymeric forms have sedimentation coefficients of 10S, 13S and 15S and are stabilized by inter-monomer disulphide bonds and by J chain. Two subclasses of IgA have been described, IgA1 and IgA2, on the basis of antigenic differences in the α heavy chain. IgA1 is the predominant subclass in serum (80–90%) whereas in secretions (see below) the subclasses are present in approximately equal proportions. The hinge region of IgA1 is approximately twice that in IgA2 and the latter class exists in two allotypic forms $A_2m(1)$ and $A_2m(2)$. As shown in Table 5.3, the $A_2m(1)$ allotype has an unusual structural feature in that the molecule lacks disulphide bonds linking the α chains with the light chains whilst these are disulphide bonded to each other. The molecule is presumably stabilized by non-covalent bonds between the heavy and light chains.

Another form of IgA, called secretory or sIgA, exists in human external secretions such as saliva, tears, intestinal secretions and colostrum where it is the major immunoglobulin and the IgG:IgA ratio is less than one. In internal secretions (e.g. synovial, amniotic, pleural and cerebrospinal fluids) the ratio is 5:1 (i.e. similar to that in serum) and furthermore, the IgA is not of the secretory type. The structure of secretory IgA is significantly different from serum IgA in that it is a dimeric molecule, with one J chain (M_r 15 000 and synthesized by plasma cells) and an additional antigenically-distinct molecule called the secretory component (M_r 70 000 and synthesized by the epithelial cells). The J chain is bound to the IgA monomers by disulphide bonds and the secretory component is bound by both disulphide bonds and non-covalent forces in the Fc regions of the IgA monomers (Fig. 5.3). The secretory component is thought to help to stabilize the sIgA dimer, to protect it from proteolytic enzymes and, as the name

Table 5.3 The subclasses of human serum IgA (after [3] with permission).

Subclass	Allotypic variants	Structure	Occurrence (%) Serum	Occurrence (%) Secretions
Ig A1	—		80	50
Ig A2	$A_2m(1)$ (90%)		20	50
	$A_2m(2)$ (10%)			

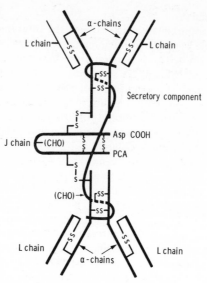

Fig. 5.3 A schematic diagram of human secretory IgA (from [1] with permission).

implies, to be involved in the process of secretion of the dimer. The secretory component is not produced by the plasma cells which secrete IgA but is produced by epithelial cells that ultimately transport the sIgA to their external surfaces and the secretory component has been viewed as an epithelial cell receptor. This process is illustrated in the lower part of Fig. 5.4 which represents the pathways by which sIgA reaches the gut (similar mechanisms occur for other mucosal surfaces). The plasma cells in the submucosa secrete IgA which is dimerized with a J chain, into the extravascular tissue space. The dimer unites with secretory component on the surface of the enterocytes, is endocytosed and transported across the cells in endocytic vesicles. Finally, the sIgA is discharged at the apical surface of the epithelial cells into the gut lumen [5]. It is important to note that monomeric IgA cannot react with secretory component.

There is a lot of current interest in the observations that antigen-committed lymphoid cells can 'home' from sites such as the Peyer's patches of the gut, via the lymph and blood, to the exocrine glands (salivary, lacrimal, mammary, urinary tract, etc.) where they produce secretory IgA as described above (see Fig. 5.5). Such migration of cells may possibly explain the observations that sIgA antibodies appear in the milk of women after intestinal antigen exposure [7].

It has been known for some time that mesenteric lymph has a much higher concentration of IgA than blood, even though the lymph delivers an enormous amount of IgA to the blood, this difference in IgA levels is maintained. Experiments in rats [8] have shown that IgA is transported rapidly from blood to bile by a process involving the hepatocytes by a mechanism which is very similar to that described for enterocytes [9, 10]. This is illustrated in the upper half of Fig. 5.4. IgA dimers (with J chain)

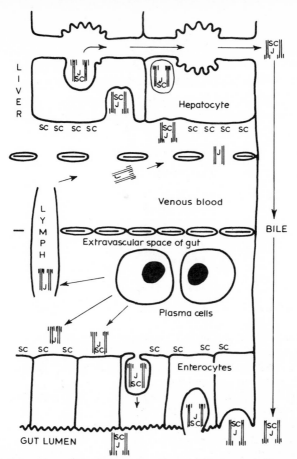

Fig. 5.4 Pathways by which sIgA reaches the gut [4]. Lower half represents the secretion of dimerized IgA by the submucosal plasma cells, the binding of the dimer to secretory component on epithelial cells, the endocytosis, transport across the cell and subsequent discharge of sIgA into the gut lumen. Upper half represents the transport of the dimer IgA via the lymphatic lacteals, the intestinal ducts and thoracic ducts to the great veins. It then reaches the portal vein and subsequently reacts with secretory component on the hepatocytes and is transported again via endocytic vesicles across the cell and is discharged into the bile and thus to the duodenum (from [4] with permission).

produced by submucosal plasma cells are collected by the lymphatic lacteals and enter the great veins via the intestinal and thoracic ducts. The dimeric IgA reaches the portal vein and combines with secretory component on the hepatocytes and the sIgA is then transported across the hepatocyte via endocytic vesicles and is secreted into the bile and hence to the duodenum. These observations provide a possible explanation for the existence of IgA dimers in blood.

As discussed earlier, sIgA is closely associated with external surfaces and it is here that it exerts its most important function. It has been said

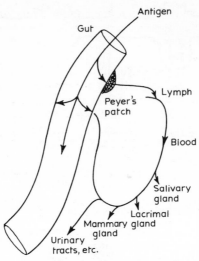

Fig. 5.5 A schematic illustration of the homing of cells from the Peyer's patches in the gut to the exocrine glands (after [6] with permission).

that sIgA can be viewed as a form of 'antiseptic paint' over mucosal surfaces because it acts there as a first-line defence against micro-organisms. Entero-pathic bacteria and viruses have to closely attach to the epithelial cells before they can infect the host and sIgA acts by steric hindrance of this attach-ment and the interaction of the secretory component with the mucus covering the mucosa may be important in achieving this. It has also been proposed that sIgA, by binding to food antigens, prevents their absorption into the blood stream and reduces the incidence of allergic reactions [11]. Furthermore, the resistance of sIgA to the activity of proteolytic enzymes in secretory fluids and the gut (a function of the secretory component?) is obviously an important property of the molecule. The multivalence of sIgA is also an important factor in the biological effectiveness of the molecule in binding to antigens because, as discussed in Chapter 3, multi-valent interactions give rise to higher functional affinity than monovalent or bivalent interactions.

Thus the structure of sIgA is uniquely adapted for its role in protecting mucosal surfaces, that is, effectively outside the body, in secretions and in the gut. In this context, it is important to remember that sIgA does not activate complement and is considered poor at promoting opsonization. These functions characterize other antibody responses in other immuno-globulin classes in serum (e.g. IgG) but would be totally inappropriate at mucosal surfaces. The functional importance of IgA has been extended from the 'antiseptic paint' concept to include the elimination of antigens which have actually penetrated the mucosa [4, 12]. Dimeric IgA in the submucosa could bind the antigen with high functional affinity, the resulting immune complex would not activate complement but would be transported to the liver via the lymph and blood (see Fig. 5.4) where it would be

74

catabolized. No allergic reaction to the invasion by the organism or antigen would thus ensue.

5.3 Immunoglobulin M

IgM, often called 19S gammaglobulin is a macroglobulin with an M_r of 970 000. Reduction and alkylation of the molecule yields stable subunits and it has been shown that the IgM molecule consists of five 7S subunits linked by disulphide bonds. Each molecule also contains one J chain which is disulphide-bonded to two μ chains. Thus the complete molecule consists of ten μ chains, ten light chains and one J chain. The μ chain has five domains, V_H, $C_\mu 1$, $C_\mu 2$, $C_\mu 3$ and $C_\mu 4$. It also contains five oligosaccharide groups attached to asparagine residues. Electron microscopic studies of IgM proteins from various species have led to the accepted structure of 19S IgM as a circular pentameric molecule stabilized by disulphide bonds and the presence of J chains (Fig. 5.6). Monomer IgM also exists in low concentrations in normal serum but can occur in higher concentrations in certain disease states (systemic lupus erythematosus, Waldenstrom's macroglobulinaemia). However, the membranes of B cells contain antigen receptors which consist of monomeric IgM (as well as other immunoglobulin

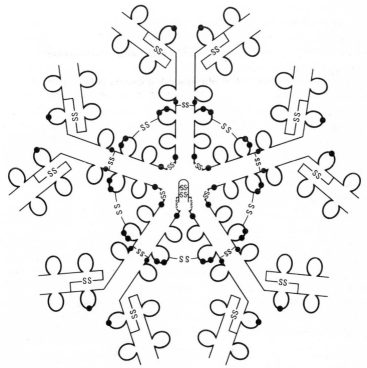

Fig. 5.6 Diagrammatic representation of human IgM. The homology regions or domains are shown together with the location of carbohydrate (●) and the J chain (from [11] with permission).

classes). In the mouse, comparison between the secreted 19S IgM (IgMs) and the membrane-bound 7S IgM (IgMm) has been made. It is clear that the IgMm differs from the IgMs only in the structure of the C-terminal end of the heavy chain (Fig. 5.7). The IgMs μ chain has an additional sequence of 20 amino acids from the end of the $C_\mu 4$ domain to the C-terminus and this 'tail' contains as its penultimate residue a cysteine which forms a disulphide bond with the J chain or with another heavy chain. In IgMm, the C-terminus of the μ chain contains a larger 'tail' of 41 additional residues within the sequence, 26 residues (569–594) are highly hydrophobic and are involved with insertion into the lipid of the B-cell membrane. Other work in the mouse has indicated that hydrophobic C-terminal regions are characteristic features of membrane-bound immuno-globulins of other isotypes. IgM antibodies are polyvalent (with ten potential binding sites for antigen) which confers on the molecules the ability to form bonds of high functional affinity with polyvalent antigens. This property, together with their effectiveness in agglutination reactions, complement fixation, cytolysis and predominant intravascular localization, indicates that IgM antibodies are particularly important in dealing with multivalent antigens such as bacteria and viruses which infect the blood-stream. It is important to recognize that a single IgM molecule can, when bound to a membrane activate the complement pathway whereas complement activation by IgG requires at least two adjacent IgG molecules on the surface. The effectiveness of IgM molecules in binding to complex antigens arises from their multivalence and the segmental flexibility of the molecule (Fig. 5.8). Electron micrographs have been taken of IgM antibodies agglut-inating *Salmonella* and these clearly illustrate this flexibility of the mole-cule [14]. In Fig. 5.9 examples are shown of the unbound IgM (a) and IgM molecules binding two flagella (b) or (c) as a 'staple' form with the binding sites interacting with one flagellum. The central discs of $(Fc\mu)_5$ of the IgM are seen in profile in the 'staple' form and are parallel to the surface of the flagellum.

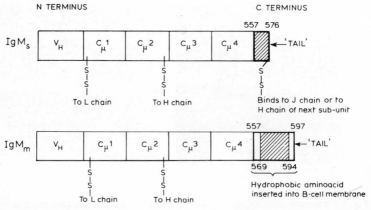

Fig. 5.7 A comparison of the heavy chains of secreted (pentameric) mouse IgM (s) and membrane-bound IgM (m).

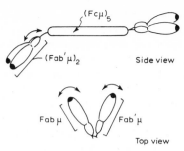

Fig. 5.8 Two possible modes of segmental flexibility of IgM. Only two of the five F(ab')$_2$ are shown (from [13] with permission).

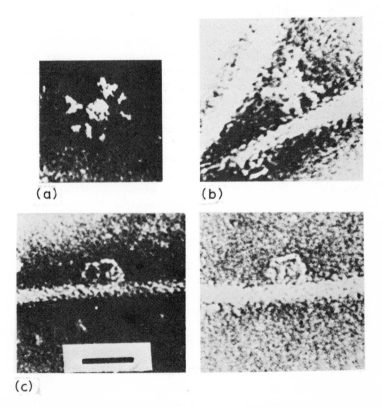

Fig. 5.9 Electron micrographs of IgM antibodies: (a) unbound IgM; (b) IgM cross-linking two bacterial flagella; and (c) two examples of IgM bound in a 'staple' form to a single flagellum (from [14] with permission).

5.4 Immunoglobulin E

As shown in Table 5.2, IgE is present in normal serum at exceedingly low concentrations. However, in individuals with parasitic infections and in certain allergic subjects, serum levels of IgE are elevated but even so, the concentration of IgE is still low. The existence of this class of immuno-globulin was discovered in 1966 [15] but detailed work on the structure of the molecule has only been made possible by the discovery of certain IgE-producing myelomas. IgE has a sedimentation coefficient of 8S in the ultracentrifuge and an M_r of 188 000 and unlike immunoglobulins M and G is heat labile. The ϵ heavy chain has four constant region domains $(C_\epsilon 1 - C_\epsilon 4)$ and a variable domain and is rich in carbohydrate. The $C_\epsilon 1$ domain has the unusual feature of two intrachain disulphide bonds and the two heavy chains of the molecule are joined by two disulphide bridges on either side of the $C_\epsilon 2$ domain (Fig. 5.10). Although present in serum at low concentrations, IgE antibodies to particular antigens (reagins) can be demonstrated by radioimmunoassays and by biological tests such as passive cutaneous anaphylaxis (PCA) or the Prausnitz–Kustner (PK) reaction. The former test is used to demonstrate IgE antibodies in experimental animals. Serum containing antibodies to a specific antigen is injected intradermally into a recipient animal and 24 hours later, after any IgE antibody has been able to bind to mast cells or basophils in the skin via the Fc region of the molecule (see below) and any non-IgE antibodies have diffused away, the animal is then injected intravenously with the antigen plus Evans blue dye. When the injected antigen reaches and reacts with the cell-bound IgE, the mast cells, or basophils, degenerate, release vasoactive amines and cause increased vascular permeability at the initial site of injection of the serum. As a result, the blue dye enters the site and the area of blueing is a measure of the amount of IgE in the serum.

The Prausnitz–Kustner test can be used to demonstrate IgE antibodies in man. Human serum containing IgE is injected intradermally into a non-allergic recipient and the subject is left for 24–48 hours during which time, the IgE antibody binds to the cells in the skin via the Fc regions and non-IgE antibodies diffuse away. The specific antigen is then injected into the *same site* in the skin, and the extent of the resulting wheal and flare reaction depends upon the presence and concentration of IgE in the serum.

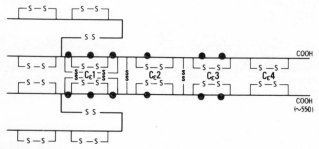

Fig. 5.10 The structure of human IgE. • = position of the oligosaccharide residues (from [1] with permission).

As mentioned earlier, the concentration of IgE in serum is very low and the current view is that the majority of the IgE in an individual is bound to the surface membranes of basophils and mast cells and it is in this form that it exerts its major biological function. However, there is evidence that the increased serum levels of IgE seen in parasitic infections are beneficial by enhancing the destruction of parasites, particularly by macrophage and eosinophil-mediated mechanisms. Furthermore, the increased vascular permeability induced by IgE mediated reactions enhances the migration of leucocytes into the site of infection.

Considerable research effort has been centred on the mechanisms by which cell-bound IgE interacts with allergen and triggers mediator release, particularly with a view to developing a rationale for the treatment of allergies [16]. It is now widely accepted that mast cells sensitized by allergen-specific IgE antibodies are triggered to release their mediators after allergen reacts with two adjacent IgE molecules on the cell surface. What is not clear, however, is the mechanism by which this interaction triggers the cell. One particularly attractive hypothesis for which there is good experimental evidence, is shown in Fig. 5.11. This hypothesis suggests that sensitized target cells bind allergen and this binding results in the generation of a signal, in the form of a pulse of positive charge, which is received by a receptor on the cell surface. The cell is then stimulated to release histamine. By the use of synthetic peptides representing human E heavy chain sequences, evidence has been obtained [16] to show that it is a small peptide in the $C_\epsilon 4$ domain of the IgE (the E-chain peptide) which provides the signal for the receptor on the mast-cell surface (Fig. 5.12).

5.5 Immunoglobulin D

This immunoglobulin was first described as a myeloma protein which did not react with antisera to the other immunoglobulin classes. In normal serum, it is present in low concentrations and is particularly sensitive to proteolytic enzymes. In addition, this immunoglobulin is heat labile. What we know of its chemical nature derives from studies on myeloma proteins. It is slightly larger than IgG, having an M_r of 170 000, it is rich in carbohydrate (12%). The number of δ heavy chain domains is likely to be four (V_H and three C_δ) but sequence data have not been reported so far. One

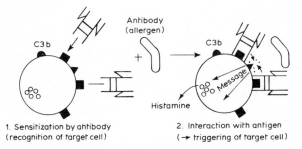

Fig. 5.11 A proposed mechanism for the triggering of mediator release following the binding of allergen by mast cell bound IgE (from [16] with permission).

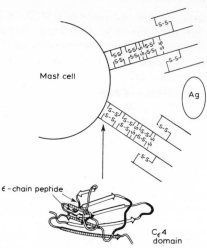

Fig. 5.12. The location in the $C_\epsilon 4$ domain of the site responsible for the signal for triggering the mast cell (the ϵ-chain peptide) (from [16] with permission).

mouse IgD monoclonal antibody has been described which has only two C_δ domains but further work is necessary to ascertain whether this is a common feature of mouse IgD. Although present in trace amounts in the serum, IgD is particularly associated with the surface of B lymphocytes where it is likely to be acting as an antigen receptor. The hinge region of IgD is large and may contribute flexibility to the molecule and hence allow the molecule to effectively cross-link antigens. During the maturation of B lymphocytes, these cells first express IgM receptors and IgM-bearing B cells are readily tolerized (i.e. made unresponsive) by antigen. IgD receptors are then expressed and these IgD-bearing B cells are more difficult to tolerize. IgD clearly plays an important role in the development of B lymphocytes.

5.6 Effector functions of antibodies

In addition to the reaction with antigen which has been discussed in Chapter 4, immunoglobulins have many other biological activities. The antigen binding activity of antibodies is associated with the variable regions at the N-terminus of both heavy and light chains. The other biological activities of antibodies are associated with the Fc region of the heavy chains of certain immunoglobulin molecules. Some of these properties are only expressed when the antibody reacts with antigen. It has been suggested that this reaction results in an 'opening up' of the hinge region of the antibody molecule and initiates conformational changes in the Fc region which activate the biological properties characteristic for that particular immunoglobulin class or subclass. This suggestion is, however, by no means proven nor indeed universally accepted. These properties differ within the various immunoglobulin classes and subclasses.

As discussed in Chapter 3, the immunoglobulin molecule is comprised of a number of domains and there is considerable evidence to support the view that the various domains are associated with particular biological functions. Table 5.4 gives the various biological properties of human immunoglobulins, the immunoglobulin class involved and the corresponding domain thought to be associated with the particular property.

Table 5.4 Biological properties of human immunoglobulins

Property	Ig class	Domain involved
Complement activation		
Classical	IgM	C_H4
	IgG	C_H2
Alternative	IgA, IgE	C_H1
	IgG aggregates	
Binding to Fc receptors on		
Human macrophages	IgG	C_H3
Mouse macrophages	IgG	C_H2 and C_H3
Monocytes	IgG	C_H3
Neutrophils	IgG	C_H2 and C_H3
Lymphocytes	IgG, IgM	C_H2 and C_H3
Syncitiotrophoblast	IgG	C_H2 and C_H3
Binding to staphylococcal Protein A	IgG	C_H2 and C_H3
Target for rheumatoid factors	IgG, IgM, IgA	Fc
Placental passage regulation	IgG	C_H2 and/or C_H3
Catabolism regulation	All classes	C_H2 of IgG

IgG is quantitatively the most important serum immunoglobulin in the human and is distributed almost equally between the intra- and extravascular fluids. Its major function is to neutralize viruses and bacterial toxins and bind to and opsonize bacteria (i.e. enhance their phagocytosis and elimination). IgG is the only immunoglobulin able to cross the placenta in the human and is thus a major defence mechanism against infection in the early part of a baby's life. The four subclasses of IgG differ in their biological properites (Table 5.5) particularly in complement fixing ability. IgG3 is the most efficient at complement fixation followed by IgG1 then IgG2. IgG4 does not fix complement.

The classical complement pathway is a complex biological system of nine protein components (C1—C9) the first of which, when activated by immune complexes, becomes able to activate the next component in the sequence which can itself activate the next component and so on, thus producing a 'cascade' effect. The activation of this system has profound effects on biological membranes and the terminal components have the capacity to cause cell death by punching holes through the membrane on which they are fixed. As well as potentiating cell lysis, activation of the complement system also promotes chemotaxis (the attraction of poly-

Table 5.5 Biological properties of human IgG subclasses

Property	IgG1	IgG2	IgG3	IgG4	Domain involved
Classical complement pathway activation (Clq binding)	+	+	+	−	C_H2
Alternative complement pathway activation (aggregates)	+	+	+	+	C_H1
Binding to Fc receptors on					
Monocytes	+	−	+	−	C_H2 and/or C_H3
Neutrophils	+	+	+	+	
Cytotoxic K cells	+	−	+	−	
Placental syncitiotrophoblast	+	±	+	+	$C_H2 + C_H3$
Intestinal epithelial cells	+	±	+	+	
Binding to protein A	+	+	−	+	$C_H2 + C_H3$
PCA in guinea-pigs	+	−	+	+	C_H3 or Fc
Control of catabolic rate	+	+	+	+	C_H2

morphonuclear phagocytes to the site of the antibody−antigen reaction) and other aspects of inflammation such as increased vascular permeability. These properties of complement activation are possessed by breakdown or fusion products of the activated complement proteins. Another pathway of complement activation has been described, the alternative pathway, in which the first components of the classical pathway are 'bypassed' and the system is activated at the C3 level by agents such endotoxins, independently of antibody. The chemistry of this process has been studied in great detail during the last few years and the interested reader should consult the information given in references [17−19].

Although IgA is present in significant levels in the serum and has demonstrable antibody activity, it is in colostrum, saliva, tears, gastro-intestinal and respiratory tract secretions that this immunoglobulin is most important. In secretions, IgA is present as a dimer associated with the secretory component and is thought to provide an immunological barrier against micro-organisms by bathing the exposed mucosal surfaces (Section 5.2). IgA does not fix complement by the classical pathway, but aggregated IgA activates the alternative pathway.

IgM antibodies, as discussed in Chapter 4, are polyvalent which gives them a high functional affinity for multivalent antigens. This property, together with their effectiveness in agglutination, complement fixation, cytolysis, and predominant intravascular localization, indicates that IgM antibodies are particularly important in dealing with multivalent antigens such as bacteria and viruses infecting the blood stream.

Some antibody activity has been demonstrated in serum IgD such as anti-penicillin and anti-diphtheria toxoid activity but the major biological importance this immunoglobulin class is likely to have is as a B-cell receptor, as discussed above.

IgE is present in the serum in extremely low concentrations and elevated levels are characteristically found in the serum of hay fever and asthma patients. Antibodies of this class can bind firmly to mast cells independent of any reaction with antigen. If an appropriate antigen (allergen) subsequently enters the host and reacts with the tissue fixed IgE antibody, a chain of events is triggered. The mast cells degranulate, vasoactive amines (e.g. histamine) are released and subsequent clinical symptoms of immediate hypersensitivity of 'allergy' such as sneezing and urticaria are produced. Current ideas on the mechanisms involved in immediate hypersensitivity have been discussed above (Section 5.4).

In addition to participating in the activation of complement, antibodies also have several other biological properties associated with structures in the Fc region. These include regulation of the placental transfer and control of catabolism, binding to macrophages and mast cells and reactivity with rheumatoid factors (anti-immunoglobulin antibodies).

Although very young animals can make antibodies, they normally depend upon transferred immunoglobulins from their mother either at the foetal or neo-natal stage. Animals differ in the way this is achieved. Some, such as the pig, horse, sheep, goat and cow receive maternal immunoglobulins entirely in the post-natal period via the colostrum. Others such as rabbit, guinea-pig, monkey and also man receive maternal immunoglobulins by placental or yolk-sac transmission. In some species, such as the mouse, rat and dog both routes operate.

In man, IgG of all subclasses is transported across the placenta. IgA, IgM, IgD and IgE do not cross the placental barrier. The mechanism by which the Fc controls transport is not known, but it has been suggested [20] that molecules to be specifically transported are bound to specific receptors on the walls of pinocytotic vacuoles of cells of the placenta and this protects them from degradation and facilitates intercellular transport. This type of mechanism has also been proposed [21] for the regulation of catabolism of immunoglobulins. Binding of IgG to the receptors protects them from degradation — IgG not thus protected is degraded by enzymes and the protected IgG subsequently released. In this way constant levels of IgG are maintained. Again, it appears that the site which regulates immunoglobulin catabolism is also in the Fc region and it is likely to be in the C_H2 region of IgG.

Antibodies are capable of binding to macrophages in the absence of antigen. It seems that such cytophilic binding with macrophage receptors occurs via the Fc region of the antibody. Immune adherence of antigen to macrophages could thus be achieved via this cytophilic antibody which is then followed by phagocytosis. It is more likely however, that adherence and phagocytosis are mediated via opsonic antibody. In this case, antibodies (IgG1 and IgG3 subclasses) bind the antigen via their antibody binding sites and binding to the receptors of the macrophages is achieved through the specific Fc binding site. Recent evidence indicates that the monocyte binding site is in the C_H3 region of the IgG.

Rheumatoid factors (RF) are anti-immunoglobulin antibodies which react with various antigenic determinants of human IgG. Such antibodies are found in the sera of patients with rheumatoid arthritis and other conditions. It has been suggested that conformational alteration of the

IgG molecule either by aggregation or reaction with antibody results in the formation of sites in the Fc with which RF can react. However, RF also react with native IgG. Genetic antigens (Gm markers) (see Table 3.1) are frequently involved in RF reactions and these Fc antigenic specificities have been demonstrated in the pFc subfragment of IgG (the C_H3 homology region) and three further sites have been localized in the C_H2 region. The relative importance of the reaction of RF with native and aggregated IgG is the subject of considerable debate.

References

[1] Turner, M. W. (1977), in *Immunochemistry, An Advanced Textbook* (eds L. E. Glynn and M. W. Steward), John Wiley, Chichester, p. 1.

[2] *Immunology Today* (1980) Volume 1 (1).

[3] Stanworth, D. and Turner, M. W. (1978), in *Handbook of Experimental Immunology* (ed. D. M. Weir) 3rd edn, Blackwell, Oxford, p. 61.

[4] Hall, J. G. and Andrew, S. (1980), *Immunology Today*, **1**, 100.

[5] Nagura, H., Nakane, P. K. and Brown, W. R. (1979), *J. Immunol.*, **123**, 2359.

[6] Hanson, L. A. (1982), *Immunology Today*, **3**, 168.

[7] Goldblum, R. M., Ahlstedt, S., Carlsson, B., Hanson, L. A., Jodal, U., Lidin-Janson, G. and Sohl-Akerlund, A. (1975), *Nature*, **257**, 797–9.

[8] Orlans, E., Peppard, J. Reynolds, J. and Hall, J. G. (1978), *J. Exp. Med.*, **147**, 588.

[9] Birbeck, M. S. C., Cartwright, P., Hall, J. G., Orlans, E. and Peppard, J. (1979), *Immunology*, **37**, 477.

[10] Mullock, B. M., Hinton, R. M., Dobrota, M., Peppard, J. and Orlans, E. (1979), *Biochem. Biophys. Acta*, **587**, 381.

[11] Taylor, B., Norman, A. P., Orgel, H. A., Stokes, C. R., Turner, M. W. and Soothill, J. F. (1973), *Lancet*, ii, 111.

[12] Peppard, J., Orlans, E., Payne, A. W. R. and Andrew, E. (1981), *Immunology*, **42**, 83.

[13] Cathou, R. E. (1978), in *Immunoglobulins* (eds G. W. Litman and R. A. Good), Plenum Press, New York, p. 65.

[14] Feinstein, A. and Beale, D. (1978), in *Immunochemistry, An Advanced Textbook* (eds L. E. Glynn and M. W. Steward), John Wiley, Chichester, p. 263.

[15] Ishizaka, K., Ishizaka, T. and Hornbrook, M. W. (1966), *J. Immunol.*, **97**, 75.

[16] Stanworth, D. R. (1982), *Molecular Immunology*, **19**, 1245.

[17] Fearon, D. T. and Austen, K. F. (1977), in *Immunochemistry, An Advanced Textbook* (eds L. E. Glynn and M. W. Steward), John Wiley, Chichester, p. 365.

[18] Muller-Eberhardt, H. J. and Schrieber, R. D. (1980), *Advanc. Immunol.*, **29**, 2.

[19] Reid, K. B. M. and Porter, R. R. (1981), *Ann. Rev. Bioch.*, **50**, 433.

[20] Brambell, F. W. R. (1966), *Lancet*, ii, 1083.

[21] Brambell, F. W. R., Hemmings, W. A. and Morris, I. G. (1964), *Nature*, **203**, 1352.

[22] Natvig, J. B. Gaarder, P. I. and Turner, M. W. (1972), *Clin. Exp. Immunol.*, **12**, 177.

6 The control of antibody production

6.1 The synthesis and secretion of antibody molecules

Immunoglobulin synthesis is achieved by plasma cells by the normal processes of protein synthesis. Immunoglobulin heavy chains are synthesized on polyribosomes consisting of 11–18 ribosomes with an ultracentrifuge sedimentation rate of 300S, and light chains on polyribosomes of 200S consisting of 4–5 ribosomes. Light chains are rapidly released from the polyribosomes and enter an intracellular pool which is maintained at a constant size. Free H chains are not normally present in the cell. Heavy and light chain synthesis is normally balanced to result in the formation and secretion of completed molecules. The result of pulse labelling experiments with radioactive amino acids have shown that there is a gradient of radioactivity from the C-terminal to the N-terminal amino acids of both heavy and light chains which is consistent with the synthesis of the chains as a single unit from the N-terminal end.

Completed immunoglobulin molecules are held in their stable configuration by covalent and non-covalent bonds and inter- and intrachain disulphide bonds play an important part in this stabilization. It is suggested [1] that covalent interchain disulphide bond formation follows the non-covalent assembly of the chains into a stable quaternary structure. H–H and H–L disulphide bonds are normally produced and the order in which these are formed has been studied by pulse labelling of immunoglobulin secreting cells and subsequent analysis of the products by electrophoresis under dissociating conditions. Two groups of intermediates have been identified as shown in Fig. 6.1. The intermediates formed vary from species to species and within immunoglobulin classes.

The situation with the polymeric immunoglobulins IgM and IgA is particularly interesting. It appears that the precursor of IgM in a mouse myeloma is the monomeric H_2L_2 form (IgMs) and that no pentameric IgM is detectable inside the cell. Polymerization to give the pentameric IgM molecule probably occurs just before or at the time of secretion [2]. A similar situation exists with polymeric IgA.

Immunoglobulins are glycoproteins containing relatively high levels of carbohydrate (particularly IgM and IgA) but the function for these prosthetic groups has not been fully established. It is likely that the carbohydrate has a number of functions including:

(1) An important role in secretion of the molecules
(2) Making the molecules soluble
(3) Protecting the immunoglobulins from degradation.

An attractive model for the secretion of immunoglobulin has been suggested and is essentially as follows:

Immunoglobulin chains are synthesized and partially assembled on

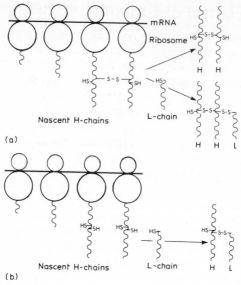

Fig. 6.1 Order of disulphide bond formation: (a) H–H before H–L; (b) H–L before H–H.

polyribosomes and carbohydrate is added to the molecule in the rough endoplasmic reticulum. The chains enter the cisternae of the endoplasmic reticulum and the four-chain molecule is formed in the cisternal space. The completed molecule is transported to the Golgi complex where further carbohydrate is added. After that it is presumed that the molecule is contained in secretory vesicles prior to secretion [3].

6.2 The genetic control of antibody biosynthesis

Our current understanding of the structure and function of genes controlling antibody synthesis is derived from three principal experimental approaches:

(1) Study of the serology and genetics of antibodies.
(2) Amino acid sequence analysis and X-ray crystallographic analysis of antibodies.
(3) The use of recombinant DNA techniques to clone genes and nucleotide sequencing of the genes.

The results of such work will be summarized in the following sections. The reader who is interested in a more comprehensive account of this exciting development in immunology should consult the references cited at the end of the chapter.

The results of amino acid sequence analysis of many human light chains has revealed the existence of two distinct regions of the chain, the variable region V_L and the constant region C_L (see above). The existence of the V_L and C_L regions in light chains has led to the unique suggestion that the well-established 'one gene–one polypeptide chain' concept of Beadle and

Tatum (1943) does not hold, and that for these polypeptides at least two genes code for one polypeptide chain [4]. Analysis of H chain amino acid sequence data revealing similar constant and variable regions has resulted in the two gene–one polypeptide chain concept also being applied to these chains. The demonstration of three homology regions in the constant part of human H chains together with the variable region has led to the suggestion that perhaps four genes are involved in H chain synthesis.

Several observations lend support to the two gene–one polypeptide chain hypothesis:

(1) The observation that V_H region rabbit allotypes are shared by IgG, IgM, IgA and IgE [5].

(2) Examples of shared idiotypy – i.e. the presence of the same variable region marker on antibodies of different H chain classes – have been described including shared idiotypic determinants on both rabbit IgG and IgM anti-*Salmonella typhi* antibodies [6].

(3) Two myelomas IgG2κ and IgMκ in the same patient but made in different plasma cells have been shown to have identical amino acid sequences up to 34 residues from the N-terminus including the hypervariable region around residue 30 (see Fig. 3.6), and also shared idiotypic determinants whilst possessing different constant regions [7].

(4) In man, a study of amino acid sequence data for both heavy and light chains suggests that separate genetic systems are involved for C and V region synthesis [8]. In light chains for example, it seems that C_λ and C_κ genes have exclusive V_λ and V_κ gene pools. Thus the single C_κ gene can be associated with any one of four V_κ subgroup genes and the two C_λ genes can be associated with any one of five V_λ subgroup genes. With heavy chains it seems that all V_H regions genes belong to a single heavy chain group, consisting of three subgroups, which can be associated with any C_H region genes. As discussed in Chapter 3, breeding studies in which allotypes have been used as constant region markers have demonstrated that the genetic loci for the constant regions of heavy chains and κ and λ light chains are not linked. The linkage of each C locus with a set of V genes means, therefore, that there are three independent gene families.

In mice one gene family (on chromosome 16) codes for the λ chain and contains V_λ and C_λ genes; a second gene family (on chromosome 6) codes for the κ chain and contains V_κ and C_κ genes; the third family (on chromosome 12) codes for the H chains and contains V_H and C_H genes. In mice, there are C_H genes for μ, δ, γ_1, γ_{2a}, γ_{2b}, γ_3, ϵ and α chains. It is clear that the number of constant genes is small, but the diversity of variable regions implied by the fact that an animal is able to produce up to 10^8 different antibodies suggests that there are far more variable region genes than constant region genes. Two theories have been developed to account for the observed diversity of antibodies.

The first, the so-called germ-line theory, suggests that a sufficient number of V genes is transmitted via the germ-line such that combination of V_L and V_H gene products provides all the necessary binding sites for combination with a large number of antigens. The second, the somatic

87

mutation theory, suggests that a small number of V genes are transmitted via the germ-line and that, during the development of the individual, V gene diversity is achieved by random somatic mutation such that immunocompetent cells each express different specificities.

The current view is that the observed variability is derived from the somatic rearrangement and mutation of *multiple* germ-line variable regions genes i.e. it is now considered that *both* the early theories are correct! Nevertheless, however they are derived, the variable region genes have to be associated with constant regions genes in order that complete polypeptide chains may be synthesized. According to the Dryer and Bennett hypothesis [4] the V region and C region genes are separate in the germline and the process of bringing them together occurs during lymphocyte differentiation. This hypothesis, which at the time it was proposed was highly radical and by no means welcomed, has been elegantly confirmed by Tonegawa and colleagues [9]. In these classical experiments, mRNA specific for the κ light chain was isolated from the mouse myeloma MOPC 321 (which contains high levels of mRNA) and radiolabelled. One mRNA sample was partially digested to serve as a probe for C region sequences only and another (untreated) served as a probe for V + C region sequences. DNA was isolated from either mouse embryo cells or from myeloma cells and digested with a restriction endonuclease. The fragments thus produced were separated electrophoretically and the fragments which hybridized with the radiolabelled mRNA probes were identified (hybridization indicates the presence of DNA sequences complementary to the mRNA probes).

The results obtained are shown diagrammatically in Fig. 6.2 with the DNA from the embryo, (Fig. 6.2a) the labelled mRNA probe for the C region (solid line) hybridized with a DNA fragment of one size whereas the mRNA probe for V + C (dashed line) hybridized with a DNA fragment of this size and also to another, smaller DNA fragment implying that the V and C region sequences are expressed of different sections of DNA. On the other hand, the two labelled mRNA probes hybridized with myeloma DNA fragments of one size only (Fig. 6.2b). These results indicate the V and C region gene sequences in the myeloma are present on the same (or similar sized) regions of DNA, thus supporting the concept that in embryonic cells, the arrangement of light chain genes is different to that in mature antibody producing cells. Thus, V and C region genes, separate in the embryonic DNA are brought together during the differentiation of the antibodyproducing cell, i.e. somatic cells can rearrange the inherited genetic information.

Immunoglobulin polypeptide chains are encoded in multiple gene segments. The DNA sequences which are translated into amino acids are called *exons* separated by intervening, untranslated, nucleotide sequences called *introns*. The DNA for the constant region of the mouse heavy chain is split into exons by a number of intervening introns. Evidence from DNA sequence analyses shows that each exon corresponds to a C_H domain. Thus for the C region of mouse IgG1 heavy chains there are four exons: one each for the $C_H 1$, $C_H 2$ and $C_H 3$ domains and one for the hinge region

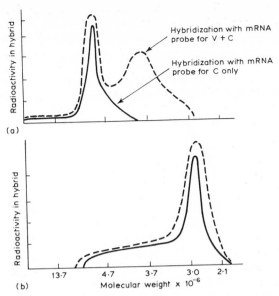

Fig. 6.2 The hybridization of labelled mRNA probes for C regions (———) and V + C regions (– – –) of κ light chains with different sized fragments of DNA from (a) embryo cells or (b) myeloma cells (after [9]).

(see Section 6.2.3). The existence of split genes for immunoglobulins are not unique but are common in the genomes of eukaryotes and viruses.

In order to produce the appropriate mRNA for translation into the polypeptide chain, the RNA transcript of the multiple gene segments (split genes) is modified by the excision of introns and the splicing together of the exons.

6.2.1 *Mouse λ light chains*

The simplest immunoglobulin chain is the λ light chain. The DNA for this chain consists of a leader sequence which codes for a hydrophobic sequence of approximately 20 amino acids (required for passage of the chain through the endoplasmic reticulum); a V_λ segment (coding for amino acids 1–97 of the variable region); a J (or joining) segment which codes for the remaining 13 amino acids of the variable region (110 amino acids) and a C_λ segment coding for the constant region. The J_λ and C_λ gene segments are in the same restriction enzyme fragment but are separated by approximately 1200 bases. Thus although in myeloma DNA the J and V segments are contiguous, a complete VC gene does not exist at the level of DNA – the V_λ and C_λ regions in DNA are 1 200 bases apart. The rearrangement of these gene segments during cell differentiation is shown in Fig. 6.3. During the maturation of lymphocytes, there is translocation in the DNA which results in the V_λ gene being aligned with the J_λ gene. The J_λ–C_λ intron (1200 neuclotides) remains intact at this stage. The transcribed RNA is

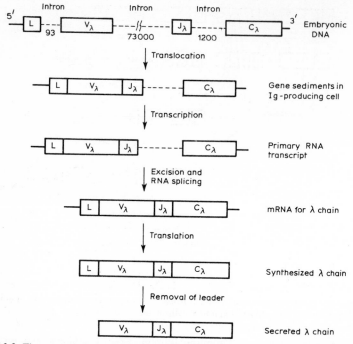

Fig. 6.3 The rearrangement of embryonic DNA and the excision and splicing of mRNA for the synthesis of a mouse λ light chain.

spliced in the nucleus and this results in the production of mRNA in which the V_λ, J_λ and C_λ sequences are contiguous and from which the complete chain is synthesized. The control of mouse λ chains is exerted by two V_λ, four J_λ and four C_λ genes (but $C_{\lambda 4}$ and $J_{\lambda 4}$ are probably not expressed).

6.2.2 *Mouse κ light chains*

In contrast to the relative simplicity of the genetic control of λ chain synthesis, the control of κ chains is far more complex. There are hundreds of V_κ genes (shown as $V_\kappa^{n-5} - V_\kappa^n$ in Fig. 6.4), five J_κ ($J_\kappa 3$ is not active) and one C_κ. Each J_κ can interact with the C_κ gene and any V_κ gene. The assembly of an active gene for κ light chains shown in Fig. 6.4 proceeds as follows. A V_κ gene with its leader sequence (L) (e.g. V_κ^{n-1}) is recombined with one of the J_κ segments (e.g. J_4). Introns remain between the L and V_κ^{n-1} exons and between exons coding for J_4 and J_5 and between J_5 and C_κ. The DNA is transcribed into a primary RNA transcript and, following excision and splicing, a complete mRNA is produced which is translated into protein.

6.2.3 *Mouse heavy chains*

The principles governing the synthesis of heavy chains are similar to those governing the synthesis of κ and λ light chains. There are thus constant

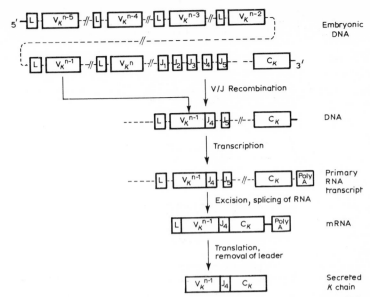

Fig. 6.4 The rearrangement of embryonic DNA and the excision and splicing of mRNA for the synthesis of a mouse κ light chain.

region genes (one for each immunoglobulin class and subclass, making eight in all), multiple V_H genes, four J_H segments and in addition, the variable region has a third controlling gene segment the D segment (for diversity). When the gene for a heavy chain variable region was analysed a sequence of 13 nucleotides at the V—J junction could not be accounted for by the V + J genes in the embryonic DNA [10]. It was argued that this segment came from the same embryonic DNA as the D gene. It is now clear that the D segments encode for the majority of the third hypervariable region of the heavy chain between the V and J regions, and code for between 3 and 14 amino acids.

Associated with each C_H gene is an additional exon M which is involved in the production of membrane-bound immunoglobulins. There is a stop codon in the C_H gene which results in the production of secreted immunoglobulin. If this codon is spliced out, the M segment is read and a membrane-bound protein is produced. The generation of an active H chain gene proceeds as shown in Fig. 6.5. The V gene is first translocated adjacent to one of the many D segments and the V—D recombinant is then translocated to a position in contact with a J_H segment. The intron separating the J of the V—D—J complex from the C_μ gene remains and is only removed after transcription and RNA splicing. Each V—D—J complex can be recombined to each of the C_H genes in sequence from the 5′ to the 3′ end of the DNA. The completed heavy chain is synthesized after the production of the complete mRNA.

The constant region genes in Fig. 6.5 are represented as a single sequence plus the exon required for the synthesis of the membrane-bound chain.

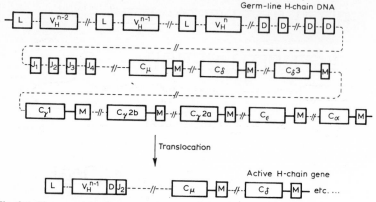

Fig. 6.5 The assembly of an active H-chain gene.

However, it has been shown that each C segment consists of four or five exons separated by introns [11]. For example, the constant region of an IgG1 myeloma protein is coded by four non-contiguous segments separated by three introns (Fig. 6.6); three of the exons code for the C_H1, C_H2 and C_H3 domains, the fourth codes for the hinge region.

6.2.4 *Recombination signals for V–J and V–D–J joining*

It is clear from what has been discussed so far that the translocation of V genes to sites adjacent to either J segments or D segments is of vital importance in generating active genes. The question obviously arises at this point as to the nature of the signals which prompt these translocations. Sequence analysis has shown that when the flanking sequences on the $3'$ side of V_κ and V_λ gene segments are compared with those on the $5'$ side of the J_κ and J_λ, there were two short, highly conserved regions of nucleotides. Each region has a stretch of nine nucleotides (nonamer) consisting of a high proportion of A and T, followed at an interval of either 12 or 23 nucleotides, by a heptamer sequence (Fig. 6.7). These two sequences (nonamer and heptamer) can form, by complementary base pairing, an inverted repeat stem structure which acts as a signal or recognition site for the deletion of DNA to bring the two gene segments together. A similar mechanism is involved in the generation of H-chain genes and includes also a recombination signal for the D region sequence. Recombination only

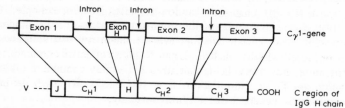

Fig. 6.6 The order of exons and introns in the segment of DNA coding for a mouse IgG1 heavy chain. Exons 1, 2, 3 correspond to C_H1, C_H2 and C_H3 respectively. Exon H codes for the hinge region (H).

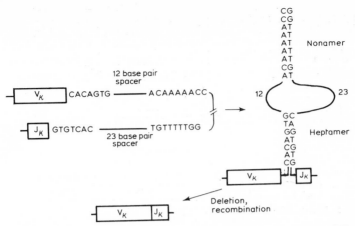

Fig. 6.7 Recognition sequences for recombination signals for the joining of V_K and J_K segments. Formation of the inverted repeat stem structures follows base pairing of the sequences separated by the 12 and 23 spacers. Only one strand is illustrated.

occurs when one of the two signals has a long spacer and the second, a short spacer.

6.3 The generation of antibody diversity

It can be estimated that the number of antibody molecules which can be synthesized by an individual is of the order of 10^8 or more. We are thus discussing here the genetics of the mechanism whereby an individual can generate 10^8 or more combining sites — i.e. the generation of V region diversity. Theories to account for — region diversity have been proposed.

Germ-line Theory: All variable region genes for the antibodies an individual can produce are inherited in the germ-line.

Somatic Mutation Theory: A limited number of germ-line V genes are diversified in somatic cells by point mutations.

There is no doubt that V region genes are inherited and evidence is available supporting the contribution of somatic mutation to the generation of antibody diversity. Thus it appears that both processes are involved. V region variability could arise in the following ways:

(1) *Multiple pairing of H and L chains*. Since the combining site is made up of both H and L chains, the random association of, for example, 10^4 H and 10^4 L chains could potentially produce 10^8 antibody combining sites. However, it is not certain whether every L chain can pair with every H chain or indeed whether each H–L pair forms an effective combining site. Furthermore, not every H–L pair need necessarily produce a different antibody specificity. Nevertheless, it is very likely that such pairing is important in the generation of antibody diversity.

(2) *Insertion of minigenes*. It has been suggested that V region diversity could be achieved by a mechanism involving the insertion of minigenes into the V region framework to generate hypervariable regions.

(3) *Multiple germ-line genes coding for V, D and J.* Amino acid sequence analysis and DNA hybridization studies have shown that there are multiple germ-line genes coding for V, D and J gene segments [12].

(4) *Multiple combinations of V, J and D gene segments.* It appears that V segments can join with many J segments, and several V segments can join with each J. Furthermore, for H chains, one V_H can join with several D segments and many D segments can pair with a J_H segment. These multiple combinations amplify the diversity in the 3rd hypervariable region of both light and heavy chains.

(5) *Variations in the joining of V and J segments.* Different nucleotide sequences coding for the amino acid 96 in the 3rd hypervariable region of the κ light chain can be generated by variations in the crossover point at which the V and J sequences combine (Fig. 6.8). Such junctional sequence diversity may also occur in the H chain system.

(6) *Somatic point mutations in the assembled V gene.* This process occurs in B cells and generates multiple nucleotide substitutions in the whole V region (both framework and hypervariable regions) of fully assembled genes. Individual nucleotides are changed and considerable evidence suggests that such mutations occur with a high frequency, particularly in IgG and IgA producing cells. The mechanism for this somatic mutation is unclear but it has been suggested that nucleotides are replaced by an error-prone repair process and that during clonal expansion of antibody-producing cells, many repair errors are introduced into the DNA.

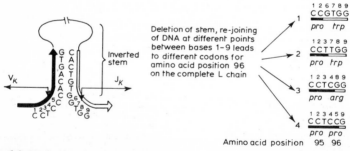

Fig. 6.8 Mechanism for the somatic generation of diversity in the 3rd hypervariable region (amino acid 96) derived combination of V_κ and J_κ sequences (after [13]).

This discussion of the genetic control of antibody production is meant as a general introduction only. The interested reader should consult references 9–19 for further, more detailed information.

References

[1] Bevan, M. J., Parkhouse, R. M. E., Williamson, A. R. and Askonas, B. A. (1972), *Progress in Biophysics and Molecular Biology*, **25**, 133.

[2] Parkhouse, R. M. E. and Askonas, B. A. (1969), *Biochem. J.*, **115**, 163.

[3] Swenson, R. M. and Kern, M. (1967), *J. Biochem.*, **242**, 3242.

[4] Dreyer, W. J. and Bennett, J. C. (1965), *Proc. Nat. Acad. Sci. (USA)*, **54**, 864.

[5] Todd, C. W. (1963), *Biochem. Biophys. Res. Comm.*, **11**, 170.

[6] Oudin, J. and Michel, M. (1969), *J. Exp. Med.*, **130**, 619.

[7] Fudenberg, H. H., Pink, J. R. L., Stites, D. C. and Waney, A. C. (1972), *Basic Immunogenetics*, Oxford University Press, Oxford.

[8] Pink, J. R. L., Wang, A. C. and Fudenberg, H. H. (1971), *Ann. Rev. Med.*, **22**, 145.

[9] Tonegawa, S. (1976), *Cold Spring Harbor Symp. Quant. Biol.*, **41**, 877.

[10] Early, P., Huang, H., Davis, M., Calame, K. and Hood, L. (1980), *Cell*, **19**, 981.

[11] Sakano, H., Rogers, J. H., Hüppi, K. *et al.* (1979), *Nature*, **277**, 627.

[12] Gearhart, P. J. (1982), *Immunology Today*, **3**, 107.

[13] Gottlieb, P. D. (1980), *Molec. Immunol.*, **17**, 1423.

[14] Adams, J. M. (1980), *Immunology Today*, **1**, 10.

[15] Leder, P. (1982), *Scientific American*, **246**, 72.

[16] Robertson, M. (1982), *Nature*, **297**, 184.

[17] Honjo, T. (1982), *Immunology Today*, **3**, 214.

[18] Williamson, A. R. (1982), *Immunology Today*, **3**, 68.

[19] Coleclough, C. (1983), *Nature*, **303**, 23.

Index

Adjuvants, 13
Affinity chromatography, 15, 17
Affinity labelling, 27
   reagents, 27
Agglutination reactions, 7, 42, 44
Allotypes, 30, 32
Antibodies
   effector functions, 80–4
   polyfunctional, 28, 65
Antibody affinity, 39, 43, 45
   biological significance, 58–60
   control, 55
   equations for calculation, 45–9
   functional, 58
   genetic control, 56
   heterogeneity, 48, 49
   immunopathological significance, 60–1
   intrinsic, 58
   maturation, 56, 57
   methods for measurement, 49–54
Antibody–antigen interaction
   cross-reaction, 61–4

hydrogen bonding, 37
hydrophobic interaction, 37, 38
ionic interaction, 37, 38
kinetics, 54
measurement, 39, 40
primary, 40, 41
secondary, 40, 41
specificity, 61–4
steric factor, 37, 38
tertiary, 40, 41
thermodynamics, 43
Van der Waals forces, 37, 38
Antibody avidity, 45
Antibody diversity
   generation, 93–5
Antibody heterogeneity, 9
   restricted, 11
Antibody response
   primary, 12
   secondary, 12
Antibody synthesis
   genetic control, 86–90
Antigen-binding cleft, 26
Ascitic fluid, 20

Bacteriolysis, 7
Bence Jones proteins, 24

Complementarity-determining regions
(CDR), 26
Coombs test, 43, 44

D segment, 91–2
Domains
    biological functions, 81

Electrophoresis, 8
Electrophoretic mobility
    of Ig classes, 9
Exons, 89–93

Germ-line theory, 88, 94

HAT medium, 18
'Homing' of lymphocytes, 72–4
Hybridization, 88–90
Hybridomas, 10, 15, 18, 19
Hypervariable regions, 26, 28

Idiotype network, 35
Idiotypes, 34
IgA, 71
    secretory, 71–4
    subclasses, 71
IgD
    function, 80
    structure, 79
IgE
    cell bound, 78–80
    cell bound, triggering of mediator
        release, 79–80
    in parasite infections, 79
    structure, 78
IgG
    enzyme digestion, 21, 23
    four chain structure, 21, 22
    hinge region, 22
    segmental flexibility, 22, 25
    subclasses, 69–70
        function, 82
    Y-shape structure, 22, 24
IgM
    monomer, 76
    segmental flexibility, 77
    serum, structure, 75–6
    valence, 57
Immunoglobulins
    allotypes, 30, 32, 33, 34
    biological properties, 81
    catabolic rates, 67–8
    catabolism, 83
    domain interactions, 29, 31
    domains, 28, 29, 31

fold, 29, 30
function of carbohydrate, 33
homogeneous, 9
idiotypes, 34
isolation, 13, 14
metabolic characteristics, 68
physicochemical characteristics, 68
placental transfer, 83
synthesis, 85–6
synthetic rates, 67–8
variants, 32
Immunoglobulin classes
    isolation, 16, 17
Immunoadsorbents, 15
Introns, 89–93

J chain
    IgA, 67, 71
    IgM, 67, 75
J segment, 90–3

Light chains
    constant regions, 24, 25
    kappa, 22
    kappa: lambda ratio, 23
    lambda, 22
    variable regions, 24, 25
    variable region sub-groups, 26, 30
Lymphocyte receptors
    affinity, 56–7

M-segment, 92
Monoclonal antibodies 10, 15
Myeloma, 9

Opsonization, 8

Phagocytosis, 8
Polyfunctional antibodies, 28, 65
Precipitation reactions, 7, 41
    fractional, 14
Precipitin reaction
    quantitative, 42

Primary Binding Tests, 44

Recombination signals, 93

Secretory component, 71
Serum antibodies
    induction, 12
Somatic mutation theory, 88, 94
Split genes, 89–90

Vaccination, 7
Variolation, 7

X-ray crystallography, 28